ABOUT ISLAND PRESS

Island Press is the only nonprofit organization in the United States whose principal purpose is the publication of books on environmental issues and natural resource management. We provide solutions-oriented information to professionals, public officials, business and community leaders, and concerned citizens who are shaping responses to environmental problems.

In 2001, Island Press celebrates its seventeenth anniversary as the leading provider of timely and practical books that take a multidisciplinary approach to critical environmental concerns. Our growing list of titles reflects our commitment to bringing the best of an expanding body of literature to the environmental community throughout North America and the world.

Support for Island Press is provided by The Bullitt Foundation, The Mary Flagler Cary Charitable Trust, The Nathan Cummings Foundation, Geraldine R. Dodge Foundation, Doris Duke Charitable Foundation, The Charles Engelhard Foundation, The Ford Foundation, The George Gund Foundation, The Vira I. Heinz Endowment, The William and Flora Hewlett Foundation, W. Alton Jones Foundation, The John D. and Catherine T. MacArthur Foundation, The Andrew W. Mellon Foundation, The Charles Stewart Mott Foundation, The Curtis and Edith Munson Foundation, National Fish and Wildlife Foundation, The New-Land Foundation, Oak Foundation, The Overbrook Foundation, The David and Lucile Packard Foundation, The Pew Charitable Trusts, Rockefeller Brothers Fund, The Winslow Foundation, and other generous donors.

ABOUT THE PACIFIC FOREST TRUST

The Pacific Forest Trust was founded in 1993 by concerned landowners, foresters, conservationists, and some of the nation's most experienced land protection experts to enhance, restore, and preserve private, productive forests, with a primary focus on the Pacific Northwest. With offices in California and Washington, PFT is: a specialized land trust for working forestlands; an information, education, and research center for stewardship forestry; and a policy institute promoting incentives for long-term forest stewardship.

A collaborative, problem-solving organization, PFT works with landowners, forest managers, public agencies, local communities, and others to sustain private forestlands for the wealth of goods and services they provide.

The Pacific Forest Trust believes that maintaining long-term, ecologically based productivity is the key to forest preservation. Private forests will be preserved only if they remain productive, and can continue to produce only if they are preserved.

America's Private Forests

America's Private Forests

Status and Stewardship

Constance Best and Laurie A. Wayburn

ISLAND PRESS
Washington • Covelo • London

Library of Congress Cataloging-in-Publication Data

Best, Constance.
    America's private forests : status and stewardship / by Constance
Best, Laurie A. Wayburn.
        p.   cm.
    Includes bibliographical references (p.   ).
    ISBN 1-55963-900-8 (cloth : alk. paper)—ISBN 1-55963-901-6
(paper : alk. paper)
        1. Forests and forestry—United States. 2. Forest conservation—
United States. 3. Forest management—United States. 4. Forest
policy—United States. I. Wayburn, Laurie A. II. Title.
SD143 .B48 2001
333.75'16'0973—dc21

                                                    00-013087

British Library Cataloguing in Publication data available.

Printed on recycled, acid-free paper   ✪
Manufactured in the United States of America
10 9 8 7 6 5 4 3 2 1

🌿 Contents

🍂 *Tables and Figures*

Tables

Figures

🍂 *Foreword*

In 1995 and 1996 the Seventh American Forest Congress brought together an unprecedented number of concerned citizens to talk about the future of America's forests. Before these meetings took place, many saw the major issues in American forest policy as springing from the acrimonious dialogue between environmental groups and timber interests over the future of national forests. After the congress, most participants realized more clearly than before that the fate of America's forests will be determined mainly by what happens in private forests, and that the political will to determine whether the owners of private forests will be engaged and assisted, lies mostly with urban people. It was clear that both private forest owners and urban citizens had been badly underrepresented in the dialogue on America's forests. Perhaps particularly underrepresented were the owners of private forestland who did not own wood-processing facilities.

This somewhat statistically contrived group, composed of what are known in the professional literature as nonindustrial private forestland owners or NIPFs, owns 60% or so of America's forests. Bound together only by forest ownership, NIPFs will by their actions determine the fate of more than half of America's forests. Add to this that they own much of the forestland threatened by urban sprawl, much that serves as urban watershed protection, and much of the most productive timberland. They now supply more than half of the industrial wood requirement of the United States, the world's largest forest products producing and consuming nation. Obviously, if only because of their forests, they deserve far more national attention than they get. Add to this their role, potential and actual, in the conservation of biological diversity, and a startling picture of past complacency and attention deficit emerges. Most biological diversity resides not in national forests or parks but on these very forests, representing virtually every forest and habitat type in the country.

Thus it is strange that a book like this one has not been written before, but wonderful that it now exists. The authors treat all classes of private ownership: industrial, nonindustrial, and tribal. The latter category is often misunderstood as federal ownership. In fact, tribal forests are private lands, held in trust by the federal government for tribal members. The tribes are increasingly active in setting direction for their forests and directly managing them. The Indian Forest Management Assessment, conducted at the request of Congress by the Intertribal Timber Council, found many positive examples of creative ecosystem management in tribal forests. Many examples of extraordinary stewardship are found as well in other nonindustrial private forests. Thus the conventional wisdom that industrial and public forests are well managed and that tribal and nonindustrial forests are not is clearly a weak generalization, and this book makes that clear.

The systematic inventory of the nature of private forests and threats to their continued well-being presented here would be valuable by itself. But this book also presents a careful inventory of the conservation toolbox available to enhance conservation in private forests. The emphasis on financial mechanisms and markets is timely and new. As the limits of top-down regulation become clearer, the need for creative, market-based approaches becomes intense.

Finally, the book describes what would constitute success in the pursuit of forest conservation, and prescribes strategies to pursue it. Its crowning glory is that the authors clearly see that maintaining forests as forests is the first conservation task required with respect to private and all other forests. If this task is neglected while we argue about precisely what kinds of forests we would like to have and keep, society itself will lose. The people will lose their forests, along with the biodiversity, water, and wood that are now provided in abundance and at low cost. That must not be allowed to happen. The challenge for the authors and all readers of this useful book is to enlarge its readership, and to build far greater understanding of the value of private forests to all of us.

John Gordon
Pinchot Professor of Forestry at Yale University
Private Forest Owner
Hollow Hill Farm
Holderness, New Hampshire
August 31, 2000

🍂 *Preface*

This book grew out of a report by the authors that was commissioned by a group of major foundations concerned with the conservation and restoration of biological resources in private forests in the United States. Known as the Consultative Group on Biological Diversity (CGBD), it sought, through its Private Forests Working Group, a better understanding of the condition of and trends affecting private forests, as well as guidance for its grant-making programs directed toward the conservation of these vital resources. In preparing revisions for this book, we widened the scope of our recommendations for a road map to use in expanding the conservation of threatened private forests. We have also updated the data to reflect those recently released from the USDA Forest Service and other sources.

The authors are the founders and leaders of the Pacific Forest Trust (PFT), a nonprofit organization dedicated to the preservation of private productive forestlands. Established in 1993, the PFT is headquartered in Santa Rosa, California, with offices in Seattle, Washington, and Boonville, California. PFT works with landowners, foresters, public agencies, and communities to develop and implement effective forest stewardship and conservation initiatives primarily in the private forests of Washington, Oregon, and California. Key to PFT's work is identifying ways to monetize forest ecosystem services and conservation, providing new financial returns to those who maintain fully functional forests. PFT acquires and manages conservation easements that protect forests from conversion to other uses and guide sustainable forest management. The organization supplies other forest conservation services, including conservation-oriented management. PFT also provides information, holds educational programs, and conducts research on stewardship forestry. Its public policy work centers on promoting incentives for long-term forest stewardship.

For further information on the programs of the Pacific Forest Trust, contact the organization at 416 Aviation Boulevard, Suite A, Santa Rosa, CA 95403; or its Washington office at 157 Yesler Way, Suite 419, Seattle, WA 98104. Visit PFT's Web site at www.pacificforest.org, or send an inquiry by e-mail to *pft@pacificforest.org.*

❦ *Acknowledgments*

This book would not have been possible without the inspiration, funding, and other support of the following members of CGBD's Private Forest Working Group: the Doris Duke Charitable Foundation, the Ford Foundation, the Great Lakes Protection Foundation, the Gund Family Foundation, the Merck Family Fund, the Moriah Foundation, the Rockefeller Brothers Fund, and the Surdna Foundation.

We were pleased to collaborate with our contributing authors, R. Neil Sampson and Lester A. DeCoster of The Sampson Group. They provided a background paper that formed substantial portions of chapter 2 and the section on public programs in chapter 4.

We would like to thank Catherine Mater of Mater Engineering in Corvallis, Oregon, who conducted a series of seventy-six interviews as background for this book. Those interviewed included fifty nonindustrial private forestland owners, thirteen consulting foresters, and thirteen state agency personnel, selected from thirteen states distributed across the forest regions of the United States.

Our profound thanks go to the members of the Advisory Group to the original report for their exceptional assistance. Their insights, questions, corrections, and generous contributions of both data and experience were essential to shaping the findings of this book. The Advisory Group includes Carlton Owen, Vice President–Forest Policy, Champion International Corporation; Charles H. Collins, Managing Director, The Forestland Group; Nancy Budge, Director of Stewardship, Mendocino Redwood Company; Peter Parker; Peter Stein, Lyme Timber Company; Russ Richardson, Appalachian Investments; Walter Sedgwick; Gil Livingston, Vice President of Land Conservation, Vermont Land Trust; Jean Hocker, President, Land Trust Alliance; Keith Ross, Vice President, New England Forestry Foundation; Kevin McGorty, Red Hills Program Director, Tall Timbers Research;

Thomas Duffus, Director of Conservation Programs, The Nature Conservancy; Ed Backus, Vice President, Programs, Ecotrust; Henry Carey, President, Forest Trust; Mike Jenkins, MACED; Bill Banzhaf, Executive Vice President, Society of American Foresters; Keith Argow, President, National Woodland Owners Association; Sam Hamilton, Southeast Regional Director, U.S. Fish and Wildlife Service; Larry Payne, Director of Cooperative Forestry, USDA Forest Service; Joan Comanor, Director–Resource Conservation and Community Development Division, USDA Natural Resources Conservation Service; Thomas W. Birch, USDA Forest Service, Northeastern Forest Research Station; Jeff Romm, Professor, RIPM, University of California at Berkeley; Paul Ellefson, Professor, Forest Policy and Administration, University of Minnesota; Hooper Brooks, Program Director–Environment, Surdna Foundation; Camilla Seth, Program Officer, Surdna Foundation; Michael Conroy, Program Officer, The Ford Foundation; Michael Northrup, Program Officer, Rockefeller Brothers Fund; Peter Howell, Program Director for the Environment, Doris Duke Charitable Foundation; Eric Holst, Program Officer, Doris Duke Charitable Foundation

In addition, we are grateful to the following interviewees for their contributions of time, information, and thoughtful comments:

DR. RALPH ALIG
Research Forester and Team
  Leader
USDA Forest Service
Pacific Northwest Research
  Station
Corvallis, OR

JIM BIRKEMEIR
Forester
Timber Green Forestry
Spring Green, WI

LINCOLN BORMANN
President
Clear Cut Consulting
Elkins, PA

ANDREA COLNES
Executive Director
Northern Forest Alliance
Montpelier, VT

RICHARD Z. DONOVAN
Director
Smartwood
Jericho, VT

KENT GILGES
Director, The Forest Bank
The Nature Conservancy
Rochester, NY

JOHN GREENE
Principal Economist
USDA Forest Service
Southern Research Station
New Orleans, LA

JOHN GREIS
Team Leader
Southern Forest Resource
  Assessment
USDA Forest Service
Southern Research Station
Atlanta, GA

Tom Hatley
Campaign Director
Southern Appalachian Forest
  Coalition
Asheville, NC

Richard Haynes
Program Manager for Social and
  Economic Values Research
USDA Forest Service
Pacific Northwest Research Station
Portland, OR

Dr. Sandra S. Hodge
Research Assistant Professor
University of Missouri

Bill Horvath
Director of NACD Policy Center
National Association of
  Conservation Districts
Stevens Point, WI

Dr. Robert J. Hrubes
Principal
Natural Resource Associates
Point Richmond, CA

Mark Ritchie
President
IATP–Institute for Agriculture
  and Trade Policy
Minneapolis, MN

V. Alaric Sample
President
Pinchot Institute
Washington, DC

John F. Turner
President and CEO
The Conservation Fund
Washington, DC

We would like to also acknowledge the assistance and advice of our expert reviewers, who took the extra time and effort to comment on the draft of this book: Richard Haynes, Ralph Alig, John Greis, and Andrew Carey of the USDA Forest Service; Scott Reed, associate dean of Extended Education at Oregon State University; and Robert Hrubes of Natural Resource Associates.

Finally, we would like to acknowledge the contributions of Jason Perry, program assistant at PFT, for his tireless support in the research and preparation of the original report, as well as the facilitation of the Advisory Group. In addition, we thank Jennifer O'Donnell, communications manager at PFT, for her assistance in updating the data and preparing the manuscript for publication. Matt Kamp of The Sampson Group also provided important contributions to the research for this book.

Even with the help of so many, the authors recognize that errors and omissions remain, and these are our own.

❧ Summary

Forests cover 33% of this country. Of that total, 58%, or 430 million acres, is owned by almost 10 million private citizens, companies, or other entities. Privately owned forests therefore are key providers of a host of essential services to our society and to the planet: supplies of wood for fuel, building materials, paper, and other products; foods, medicinals, and decorative florals; diverse habitats for great numbers of plants and animals; stores of genetic wealth; watershed functions; climate stabilization and carbon sequestration; recreational opportunities; and aesthetic enjoyment for millions of people. Further, these forests are arguably the most productive in the United States, in terms of timber capacity and other measures, as compared with publicly owned ones. Private forests generate tremendous economic value and employment in the United States, as well as provide for invaluable noneconomic benefits.

Yet the viability—indeed, the very existence—of America's seemingly vast landscape of private forests is increasingly threatened by population growth, sprawling urbanization, fragmentation, and nonforest development. From the perspective of maintaining biodiversity and overall ecosystem wealth, most private forests have been impoverished over the course of U.S. history. Forest values in many parts of the country are accelerating their decline relative to returns available from real estate development or forest liquidation. While some trends in maintaining forest cover are relatively positive—with considerable reforestation of cutover regions like New England since the 1920s—many major forest states and regions have been experiencing accelerated forest loss in the last twenty years. In 1998, the National Research Council (NRC) reported that an additional 20 million acres of private forest nationwide is at risk of being lost by 2020. Further, based on recently released data from the Natural Resources Conservation Service (NRCS), on average almost a million acres of private forest were lost to development each year from 1992 to 1997—a 70% increase over the previous decade, 1982–1992 (NRCS 1999). North Carolina, California, Florida, Georgia, Massachusetts, and Washington lead the country in forest loss.

Forest loss is made worse by the growth of metropolitan areas, especially in the South and West, which swallow up huge tracts of forest and urbanize them. Overall, the rate of fragmentation of private forest tracts is increasing with alarming speed: on average, almost 2 million acres of forestland per year—an area about the size of Yellowstone National Park—were broken up into parcels smaller than 100 acres between 1978 and 1994. The faster turnover in forest properties—those of both industrial and other private owners—feeds into the market for smaller residential or recreational parcels.

The forest landscape is unraveling. There are indications that the ecological integrity and economic functionality of our private forests are declining as well. Unlike public forests, which are likely to remain as forest even as people argue over their management, private forests are not guaranteed to be around tomorrow. Private forest owners are free to seek economic returns that represent the "highest and best use" of their properties, taking advantage of nonforest market opportunities as they grow.

In the face of these daunting trends, there is a growing awareness of the need to protect private forests on the part of forest landowners, communities, public agencies, conservation organizations, and philanthropies. In response, a growing suite of tools is available to accelerate and expand the conservation and stewardship of U.S. private forests. Some of these tools are well established and their use simply needs to be expanded. Others are emerging initiatives or concepts that need to be considered fully and potentially invested in as promising experiments. In fact, bold new conservation ventures likely need to be undertaken to address the scale of threat to private forests.

The success of large-scale private forest conservation is predicated on engaging the self-interest of private forest owners. While many forest owners are excellent stewards of their land, without gaining an expanded commitment to forest stewardship and conservation by the community of forest owners, there is no hope of reversing the trend of increasing loss of forest ecosystem functions and extent. Forest conservation needs to make sense to forest owners and help them fulfill their goals of forest ownership. The markets for conservation need to be expanded and returns for conservation-based forest management increased. Further, effective conservation efforts require better understanding of who these owners are, what motivates them, and how to reach them effectively—especially those estimated 68,000 owners who control 45% of private forests in tracts of 500 acres or more. At the same time, there are hundreds of thousands of owners of recently created smaller forest parcels who need to become more engaged in forest stewardship to better overcome the effects of forest fragmentation. The overview of private forest ownership provided in this book provides a first step toward this better understanding.

Of perhaps equal importance is the need to gain greater public understanding of the contributions of private forests to our lives and society. The growing lack of connection between people and forests in our urbanized society is itself a key barrier to increased public and private investment in forest conservation. To dramatically reduce the simplification, fragmentation, and loss of private forests, American culture needs to more fully value forests for all their contributions. This process can begin by strategic investment in what we call cultural tools. Some that we describe help forest owners integrate stewardship and conservation into their forest management. Others make private forests more meaningful and relevant to the general public.

All of this requires a much better understanding of the condition and characteristics of privately owned forests. The lack of complete, consistent, and timely data about the suite of private forest resources—not only timber—hampers everyone's conservation and stewardship efforts. The description of private forests provided here is a beginning in the process of consolidating the all too diverse and poorly distributed data so that accurate and useful portraits of private forests and national, regional, and local trends in their condition may be compiled.

In seeking to address the many threats to private forests, it is important to remember that there are no silver bullets that will resolve the complex challenges of forest loss in one or two blows. In fact, many tools and approaches will need to be used to address different aspects of the challenge. The key is understanding which tools are best for what, and how they can be most effectively applied both individually and synergistically. This book provides a comprehensive survey of the various tools in the proverbial toolbox, organizing them within three broad categories: public programs, cultural tools, and market-based tools. Examples of ongoing projects or emerging initiatives are provided for each tool.

The final chapter details an action plan to accelerate the conservation of private forests. We argue for a sustained, collaborative effort among forest landowners, public agencies, nonprofit organizations, forest communities, and others with an interest in conservation of private forests. The conservation objectives that are the focus of this action plan are to

1. turn the tide on private forests loss;
2. dramatically reduce the fragmentation of larger forests;
3. create ways to functionally reassemble the landscape;
4. fuel the restoration of ecosystem wealth; and
5. build a culture that values forests.

While conservation and stewardship of smaller forests are important to reassembling a fragmented landscape, we believe the primary focus of conservation investment needs to be on those forests most at risk for further fragmentation and conversion, especially larger forests (with 500 acres or more) everywhere. Those larger tracts in still rural areas just beyond the leading edges of growing urban areas represent a special opportunity for efficient conservation before they fall victim to the next wave of development. Of great concern are those larger forest tracts in transition, whether through corporate restructuring or family succession. Larger private forests near recreational public lands and waterways are also highly threatened, as are larger forests with well-stocked older stands and high biodiversity value.

We recommend that private forest conservation efforts be concentrated on two basic kinds of actions: (1) bringing the conservation market to scale, for rapid gains in private forest protection; and (2) integrating conservation into forestry—and forestry into society.

The returns from conservation and stewardship need to increase in order to compete with those available from development and degradation. Expanding the philanthropic and public capital available for conservation of public benefit resources on private forests will enable larger, financially driven forest ownerships to expand their commitments to conservation. Developing new markets for ecosystem services provided by forests—such as water provision or carbon sequestration—will expand the sources of capital. By increasing such financial incentives, conservation can become more integrated into forest management's business as usual, helping stewardship become the norm. Therefore, we recommend the following major initiatives:

1. Provide new conservation capital for intervention during the turnover in ownership of significant forest properties through public, philanthropic, nonprofit, and private partnerships.

2. Expand the public market for conservation through improved and expanded funding programs that protect public benefits of private forests.

3. Catalyze the development of new sources of funding for ongoing conservation through markets for forest ecosystem services, in particular for forest-based carbon sequestration and watershed services.

4. Improve returns from long-term forest stewardship and conservation through changes in key areas of taxation.

5. Increase access to liquidity and traditional sources of capital for smaller landowners who are otherwise constrained in making investments in forest stewardship and conservation.

6. Increase returns for managing forests with high native biodiversity values.

In addition, we believe that a series of specific actions will contribute substantially to building a culture that better values forests. This side of the strategy complements the other by seeking to provide forest owners and stakeholders with the information, options, and motivation to advance private forest conservation and stewardship over the longer term. These active initiatives seek to resolve threats to private forests through communication, education, and assistance, helping forestry finish the evolution from logging to forest ecosystem stewardship. We recommend the following major initiatives:

1. Tell the story of good forestry and its financial as well as ecological returns to inspire forest owners and build public support for forestry as a desirable land use.

2. Build support among policymakers for increased public investment in private forest conservation.

3. Better identify and understand key forest landowners in order to provide them with more effective forest conservation and stewardship services.

4. Convene and build new coalitions and partnerships among natural forest conservation allies.

5. Enhance regional stewardship capacity to support landowners in their conservation efforts.

6. Improve access to useful scientific information to advance forest conservation and stewardship.

If those concerned about the fate of America's private forests can co-operatively implement a sustained effort such as we have described, both immediate and long-term gains will be made in the conservation of these crucial forests. If, on the other hand, those who value private forests fail to reach out across traditional divides, and fail to engage the wider public, it is very possible efforts will be too little and too late to forestall the acceleration of forest loss we are seeing today. Given the immense value of America's private forests—measured biologically, financially, or spiritually—we believe there is tremendous potential to mobilize broad engagement in their successful conservation. Those who already know and prize these forests must take the lead in meeting this challenge.

❧ *Introduction*

Privately owned forests are faced with increasing conversion pressure in many regions of the country. In addition, there have been long-term negative impacts to forest health and ecological functioning due to a history of settlement, logging, and other uses. These factors have led growing numbers of people, including foresters, public agencies, residents of forest regions, and others, to turn their attention to the challenge of sustaining private forests on the landscape.

Why care about the condition and future of privately owned forests? Simply, they comprise almost 60% of our nation's forests and are home to millions of people and countless other creatures, including many threatened and endangered species. They filter and regulate the flows of many key watersheds. They hold great stores of carbon. They make up the landscape that defines and embraces our public forests and protected areas. The important values of public forests cannot be adequately protected and sustained without addressing those private forests: ecosystems and creatures do not recognize legal boundaries and mapped ownerships. Yet there are no guarantees that privately owned forests will stay as forests tomorrow. The United States has already lost more than 30% of its forest area since European settlement. More is lost every day in our sprawling metropolitan areas. Through ecological degradation, fragmentation, and conversion, the essential resources provided to society by privately owned forests are being eroded and diminished. Whether the primary concern is habitat, water quality, climate stabilization, or timber supplies, there must be a concern for the future of America's private forests in the face of intense and growing pressures.

America's private forests are dramatically diverse, as is the array of private forest owners and their motivations in owning these forests. The threats to private forests are an interlocking web of environmental, economic, and social forces of daunting complexity. Although many volumes and papers have been written about aspects of our nation's private forests, none has sought to bring together these elements into one book written in a language accessible to the uninitiated reader. With this book, we hope to stimulate the interest of those who are new to the issues facing private forests as well as that of seasoned and sophisticated forest owners, managers, policymakers, and funders.

In this book we summarize the state of privately owned forests in the United States. We identify the major threats to the continued existence of these forests, as well as barriers to their conservation. We then outline the existing tools, support programs, and new initiatives or techniques in development to encourage the conservation of private forests. We close by providing an action plan of recommendations to guide collective efforts to accelerate the conservation of private forests and their native biodiversity.

This book is meant to inform people's efforts to better conserve and steward the biodiversity of privately owned forests. Although timber values are addressed, they are considered within the overall suite of goods and services that will be sustained as we improve our ability to conserve private forests.

The data on forests are organized by a sometimes confusing array of terms. The following definitions can help you to see the forest for the data.

Forestland, according to the USDA Forest Service, is land that is at least 10% stocked by forest trees of any size, including land that formerly had such tree cover and is expected to regenerate, naturally or artificially. Forestland also includes pinyon-juniper and chaparral areas in the West, as well as nonforest soils planted with trees. However, the USDA Economic Research Service and Natural Resources Conservation Service definitions are based on land use. This leads to differing estimates of forestland acres, as will be shown in this book. *Forestland* can be used synonymously with *woodland,* which sometimes refers to forestland that does not contain commercial timber species.

Timberland is forestland that is capable of producing more than 20 cubic feet per year of commercial species for processing (and therefore not including fuel-wood). This is the category of forest for which the most data are available.

Reserved forestland is timberland that is reserved from timber harvest by statute or regulation and includes federal and state parks and wilderness areas.

Conservation, as we use it, means the protection of private forests from degradation and loss to better ensure their long-term productive use and enjoyment, benefiting their owners, communities, and the public at large.

Stewardship, another word we use, means the active investment of time, money, knowledge, and other resources into the improved health and ecological well-being of forests. Stewardship expresses an ethic and approach to forest management that reflects the benefits to the public, to future generations, and to forest ecosystems that are derived from these investments by private forest owners.

The challenge of completing this book was compounded by the many inconsistencies and gaps in the data available on private forests. The sources of data on private forestlands, like the forests themselves, are diverse and dispersed. As private forests are not under federal regulation, and state forestry programs vary, there is no central depository of information about this huge expanse of our national landscape. The few federal programs that deal with private forests and their resources reside in several agencies, with differing quality and quantity of data collection. These often lack standard classifications of vegetation cover, including the definition of *forest.* According to a publication by the NRC (1998), "The information available to describe non-federal forests is often out of date, gathered by agencies with conflicting interests, and inconsistent in form and presentation, making its aggregation across regions impossible."

Following are the primary federal sources of data we used:

The Resource Planning Assessment (RPA) program of the USDA Forest Service prepares a strategic plan every five years that focuses on timber resources and projections. The RPA utilizes data provided by the Forest Inventory Analysis units and Timber Management staffs around the country and produces a major publication titled "Forest Resources of the United States." While new RPA databases and analyses based on 1997 data have just become available from the USDA, for many resources we must utilize information that is up to ten years old.

The USDA Natural Resources Conservation Service (NRCS) produces the Natural Resources Inventory (NRI) every five years. The NRI inventories nonfederal rural land cover and use, soil erosion, prime farmland, wetlands, and other resource characteristics. We utilized the recently published revised 1997 NRI.

The U.S. Department of the Interior's U.S. Geological Survey (USGS), through its Biological Resources Division, recently published

"Status and Trends of the Nation's Biological Resources," covering the abundance and health of U.S. plants, animals, and ecosystems.

In preparing this book we worked with many experts, interviewed some 120 people across the country, and combed through the literature on timber and ecological forest resources over nine months in our effort to create an overview of this complex topic from a national perspective. We were guided by an Advisory Group of twenty-seven, including seven forest landowners (both large and small, mill owners and non–mill owners); representatives of five land conservation organizations; two forestry associations; three sustainable development organizations; four public agencies; four foundations; and two academics. The Advisory Group played an important role. Its members were interviewed by the authors and were available for questions, supplied data and other sources of information, and provided invaluable review and comment on drafts of this book. One meeting of the Advisory Group was held midway through the project to review our data, test our assumptions, consider tentative conclusions, and begin the articulation of the strategic recommendations.

In discussing the condition of private forests and their resources, we did not seek to rigorously assess and quantify their status or the sustainability of their management. Indeed, available data does not support such a quantitative analysis. We must defer to the scientists and others taking part in larger and longer-term projects that address this challenge. We have, however, provided a panoramic view of the current status of our nation's private forests so as to solidly ground the ensuing discussion of threats as well as solutions. Although this book was prepared for nonscientists, we have been careful in our analyses to stay as close to the data as we can. Yet there are many forest resources for which qualitative information far outweighs quantitative. In the absence of hard data or detailed studies of issues such as the large-scale and long-term impacts of forest fragmentation on ecological integrity, we have extrapolated where it seems warranted and drawn on the expert opinion of our Advisory Group, interviewees, and reviewers.

Scientists, policymakers, and other interested parties continue to wrestle with the still unresolved definitions and measures of forest integrity and ecological functionality. One such effort is the Montreal Process, an international effort involving nations including the United States that are interested in achieving agreement on principles of conservation and sustainable forest management as well as on criteria and indicators of sustainability. There are seven criteria, each with several indicators: conservation of biological diversity; maintenance of productive capacity of forest ecosystems; maintenance of forest ecosystem health and vitality; conservation and maintenance of soil and water resources; maintenance of forest contribution to global carbon cycles; maintenance and enhancement of long-term multiple socioeconomic benefits to meet the needs of societies; and legal, institutional, and economic frameworks for forest conservation and sustainable management. The USDA Forest Service is leading the U.S. effort within the Montreal Process. Given the limitations on the data available on private U.S. forests, even this effort is compelled to utilize proxy data and extrapolations, while seeking to develop new data as feasible.

In the course of our research, we asked a series of questions of ourselves and others:

- Given that those who own America's private forests are central to the forests' conservation and stewardship, who are these owners? What are their motivations in owning forests? How have those motivations shaped the private forests of the United States? How are new trends in ownership likely to shape the forests of the future? How are trends in ownership size affecting the nature of private forests?
- Given our desire to better conserve America's private forests, what is their condition today? How have they changed over time? What social, economic, and ecological influences are shaping them? What are the impacts of development, logging, and other management actions on forest resources, both timber and nontimber?
- Given the historic and ongoing loss of forests in the United States, as well as the degradation of certain ecosystem functions, what are the drivers of these conditions? What cultural and economic barriers exist to conservation? What are landowners' perceptions of the threats to private forests and the barriers to conservation? What are the perceptions of public agency personnel who work with private forests and their owners? What are the perceptions of foresters, scientists, conservationists, and others? How are these drivers and influences related?
- Given the complex nature of the threats to private forests, what programs, initiatives, and techniques exist to address these threats and overcome barriers to private forest conservation? What current public pro-

grams have been effective is assisting landowners in their conservation and stewardship efforts? What nonprofit programs are effective? What roles are played by various service providers, such as universities, landowner and forester associations, land trusts, and others? What are the best uses for different tools and what are their limitations? What are the social conditions under which forests will be better valued and conserved?

- Given all of the above, how can those concerned deploy and leverage scarce resources to strategically accelerate the effective conservation of private forests in the United States? What forests or ownerships are most at risk? What are the goals and objectives of a national effort to advance this conservation? What new sources of capital can be mobilized to help meet these challenges?

We expect to keep working on these answers, and hope this book inspires others to join in that work. There are many more questions than are presented here, with answers that may illuminate new approaches to resolving the threats to private forests. There are great needs for additional research, development, and, above all, implementation of these conservation tools and strategies. We encourage others to build from this work to ensure that America has at least as great an area of private forests at the end of this century as exist now. Moreover, we hope the ecological integrity, health, and functioning of these forests is at an even higher level.

America's Private Forests

Part One

An Overview of America's Private Forests

THIS SECTION PROVIDES background and context for our recommendations for an action plan to promote the conservation of private forests in the United States. We will examine the changing nature of private forests and private forest ownership at the end of the twentieth century as a basis for understanding the major threats to the maintenance of this large and important resource as fully functioning forests in the twenty-first century.

Forests occupy an estimated 747 million acres, or one-third of the nation's land area, and 58% or 430 million acres are privately owned.[1] The private forests of the United States account for 30% of the world's total. U.S. forest types and forest ownership are many and varied, forming a complex whole. While much attention (and no small amount of controversy) is directed at the management of federal forests and the operations of the forest products industry, both hold relatively modest portions of the nation's total forests (figure 1). Almost half of America's forests are held by some 9.3 million nonindustrial private owners. Any serious forest conservation effort must therefore consider what is happening on these lands and why.

We will look carefully at the people and businesses involved in forest ownership. Today humans are the most significant force in ecosystems, shaping the forest as surely as fire does. It is humanity's collective sense of values that has guided our interactions with forests through history. And those values change. What was once seen as "good" (carving a field out of the wilderness for a family's home) may now be seen as "bad" (cutting the last old growth, or fragmenting a forest habitat). The values, needs, and goals of forest landowners must be central to our understanding of the opportunities and strategies for conservation.

We also look at the land itself, and the character of the forest ecosystems involved.[2] These are not the same forests encountered by explorers and pioneers, nor are they even the same as those of a half century ago. In many places, they are forests that, in an ecological sense, contain an entirely different set of species, structures, and processes than may ever

---

1. The forest statistics in this report are taken either from USDA Forest Service data for 1992 (Powell et al.) or from the recently released 1997 RPA.

2. We use the terms *forests* and *forest ecosystems* interchangeably, since we view a forest as a discrete piece of the landscape, complete with all living and nonliving components, affected by all energy and material flows into and out of it. For literary convenience, we avoid the longer term except when context seems to demand its use.

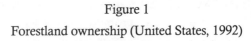

Figure 1

Forestland ownership (United States, 1992)

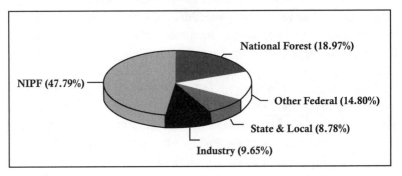

have existed before. Where that is the case, scientists may have few guides as to how these forests will continue to change and adapt in the future.

Will this combination of human and natural changes result in a sustainable forest—sustainable in the sense that the forest will continue to maintain its ecological processes and functions—or will they result in an entirely different ecosystem, perhaps one that is grassland, or even desert? While deserts have their own intrinsic values, they are not the same values forests have, and the difference is significant to many people. Further, if that difference was created, or exacerbated, by human actions, people are responsible for what has occurred. Thus our focus is on the forest and its capacity. Can it sustain its ecological integrity under the pressures of current human activity, or is its future imperiled? Can we identify the root causes of the potential threat and affect them in a more positive way?

Chapter One

Who Owns the Forest and Why?

Profile of Private Forestland Owners

The ownership of U.S. private forests is exceptionally diverse, as tables 1-1, 1-2, and 1-3 illustrate. In this chapter we will provide information that will promote understanding of the universe of owners. Those concerned with the future of private forests need to go beyond received wisdom and look closely at who owns what and why. The chapter closes with an analysis of recent trends and a discussion of implications for the future.

The data that exist to describe the nature of private forest ownership are useful to examine, but are not as complete as they could be. The primary current national source is a 1994 study led by Thomas Birch for the USDA Forest Service. This survey of private forest ownerships included in its base all of the private timberlands as well as key portions of other forests, estimating that there were around 9.9 million private ownerships holding 393 million acres of forestland (Birch 1996). The estimate of forestland acreage used by Birch is different from the usual estimate of acreage cited by the USDA Forest Service (424 million) because Birch started with the base of private timberland (358 million acres) and added to it a sampling of other forestland. Table 1-2 presents a snapshot of forest ownership by size classes of ownerships. Refer to appendix B, tables, for more ownership detail by state and region.

Table 1-1.
Forestland acres (000s) by ownership in the United States, by region

Region	All Owners	Forest Service	Other Federal	State and Local	Forest Industry	Percent Total	NIPF Total	Percent Total
Inter-mountain	138,447	71,255	26,119	5,921	2,939	2.1%	32,213	23.3%
Alaska	127,380	11,250	55,499	24,756	0	0.0%	32,213	28.1%
Pacific NW	51,612	22,352	5,011	3,663	9,595	18.6%	10,991	21.3%
Pacific SW	40,296	16,748	3,918	1,720	3,140	7.8%	14,768	36.6%
Great Plains	4,798	1,223	100	226	0	0.0%	3,248	67.7%
North Central	84,842	9,064	1,549	15,629	3,814	4.5%	54,785	64.6%
Northeast	85,484	2,544	776	11,594	11,158	13.1%	59,412	69.5%
South Central	125,438	6,870	2,859	2,954	22,529	18.0%	90,226	71.9%
Southeast	88,662	5,470	4,114	3,291	14,511	16.4%	61,277	69.1%
Total U.S.	**746,958**	**146,777**	**99,945**	**69,754**	**67,687**	**9.1%**	**362,796**	**48.6%**

Table 1-2.
Profile of U.S. private forestland ownership by owners and acres

	1–9 acres	10–99 acres	100–499 acres	500–999 acres	1,000+ acres	Total
Landowners	5,795,000	3,480,000	559,000	41,000	27,000	9,902,000
% Total owners	58.52%	35.14%	5.65%	0.41%	0.27%	100%
Forest acres	16,600,000	107,600,000	91,600,000	24,500,000	153,000,000	393,300,000
% Total acres	4.22%	27.36%	23.29%	6.23%	38.90%	100%

In addition to the Birch study, we utilize regional and subregional surveys and analyses of ownership characteristics from a variety of sources, which will be cited as we proceed. Unfortunately, there are many basic things that the data do not yet tell us, such as the median size of ownership, the average sizes of parcels within ownerships, and the income levels, races, and other demographic details of the individual owners. Nonetheless, a review of the literature presents us with very useful information for informing a strategy for conservation of private forests.

Private forests are generally separated into two categories: industrial and nonindustrial. This distinction can be somewhat confusing, since *industrial*

Table 1-3.
Forest owners by type of entity

Ownership Type	Ownerships	% Ownership	Acreage (000s)	% Acreage	Average Acreage
Forest industry	13,300	0.13%	79,715	20.26%	5994
Farm	2,431,300	24.55%	111,450	28.33%	46
Industrial business	30,700	0.31%	9,031	2.30%	294
Real estate	236,200	2.39%	15,948	4.05%	68
Other business	74,600	0.75%	3,986	1.01%	53
Recreation club	116,200	1.17%	7,768	1.97%	67
Public utility	800	0.01%	2,248	0.57%	2810
Individuals	6,765,000	68.32%	133,521	33.94%	20
Other	233,600	2.36%	29,722	7.56%	127
Total	**9,901,700**	**100.00%**	**393,389**	**100.00%**	**40**

in this context means a company that owns both forestland and mills to process forest products. Thus, industrial forestland may be held in million-acre tracts by a multinational corporation, or it may be a 160-acre parcel owned by a small local sawmill. A large oil company or institutional pension fund that owns forestland but does not own a wood-processing facility will be found in the nonindustrial category. The nonindustrial category encompasses land that may be held by an individual with 5 acres as well as hundreds of thousands of acres held by Alaska Native corporations. In some cases it is possible to tease some of these differences out of the data sets, but often the distinctions are difficult or impossible to determine.

Table 1-3, which describes ownership types compiled from the 1994 Birch study, provides some illumination on the kinds of entities that own forestland. Looking at this table, we can begin to appreciate the diversity of nonindustrial owners. While individuals and farms are the largest ownership types overall, the forest industry clearly has the most concentrated control.

It may be most useful to consider forest landowners not by distinctions of mill ownership, but rather by size. Large, medium, and small ownerships, whether with or without mills, tend to share common attributes within their size categories. Therefore, we will utilize the following categories for analysis of the characteristics of private forest landowners:

Residential Forest Owners: 1–9 Acres

This size ownership is essentially residential in nature. While these parcels retain some forest characteristics and contributions, from the point of view of ecological functionality and timber productivity they are effectively converted from forest use. Even with sporadic timber harvesting or ongoing wildlife management, very small forest tracts are dominated by residential uses, including buildings, exotic landscapes, and domestic animals. Tree management is generally more horticultural than silvicultural at this size.

Small Forest Owners: 10–99 Acres

This size ownership represents the average current tract size. This is also the size category experiencing the greatest growth in numbers of owners and acres represented. Though very fragmented in nature, these forest properties can still provide many major forest values. They can be managed for periodic timber or other forest-based revenue, though major harvest typically occurs only once or twice in an owner's lifetime.

Medium Forest Owners: 100–999 Acres

This class of forest ownership can form the building blocks of larger, more functional forest landscapes. As individual holdings, these parcels can be managed for regular economic return for forest products more readily than can parcels of smaller size. This size ownership is akin to the "shrinking middle class," contributing acres through increasing subdivision to the swelling numbers of small forest landowners in the last twenty years.

Large Forest Owners: 1000+ Acres

At this threshold, forestland becomes more likely to be held for commercial timber production and decisions regarding its management are more likely to be driven by financial considerations than is the case with smaller ownerships. Although on a national basis this class of ownership has slightly decreased in extent, this trend is very regional in nature, with the North losing more large tracts to fragmentation and the South consolidating medium-sized tracts.

As shown in table 1-2, while almost 60% of forest owners own less than 10 acres, their impact on the forest landscape is a tiny 4%. At the other end of the spectrum, large forest landowners comprise just one-quarter of 1% of the total but control almost 40% of U.S. private forests.

Figure 1-1.
Forest ownerships as percent of total by size
(including size categories > 9 acres)

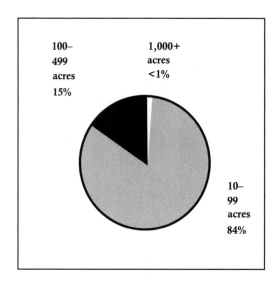

Figure 1-2.
Forestland ownerships as percent of total private forest acreage
(including size categories > 9 acres)

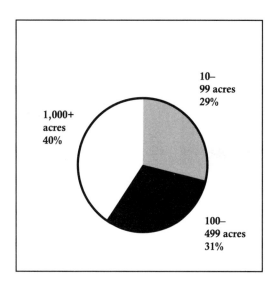

Figure 1-3.
Primary reason for owning forestland (millions of owners)

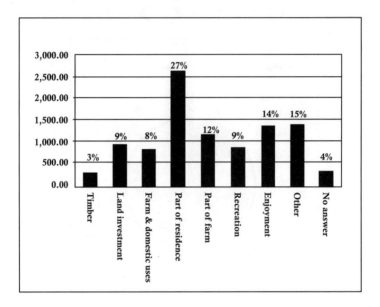

Figure 1-4.
Primary reason for owning forestland (millions of acres)

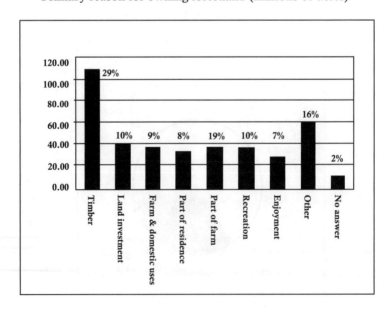

If the residential owners' share of total private forest ownerships is excluded, the ownership picture changes to better reflect the realities of forest management and conservation. With an estimated forestland base of 376,700,000 acres having 4.1 million owners, the average parcel is 92 acres (versus 40 acres if the residential owners are included). Small landowners comprise 84% of owners and control 29% of the forest. Medium landowners comprise 15% of the total with 31% of the acres. The large landowners still number less than 1% while accounting for 40% of private forest (figures 1-1 and 1-2).

Why People or Businesses Own Forestland

The following charts based on the Birch 1994 data (figures 1-3 and 1-4) give us some indicators of the reasons different entities own forests. (Because of the way the Birch data is published, these and other data presented in this section include the 5.8 million residential forest owners, thereby providing some bias toward this group.) While the data are not crystal clear, they suggest that U.S. private forests are owned roughly equally by those with primarily "productive" or economic motives and those who own forests for "nonproductive" personal, cultural, and/or ecological values.

Almost 40% of owners, by far the largest block, state that their primary reason for owning forestland is simply that it is a part of their residence or farm. Another 23% characterize their primary reason as being for recreation or for the sheer enjoyment of owning forestland.

Just 20% of forest owners state that their primary reason for ownership is economic. These owners have forests either for timber (about 3%), real estate investment, or as a productive part of a farm or home, yielding timber, fence posts, or firewood. However, as figure 1-4 illustrates, this group of owners controls almost half of U.S. private forests, with timber production alone representing 30%.

Still, substantial forest acreage, often in the smaller ownership size classes, is held for its noncommercial values. In fact, included in the 16% of "other" uses is cultural use by Native Americans. (Other uses also include mineral extraction; for owners of mineral rights, the trees are incidental to other economic use.) Various surveys of forest landowners indicate that smaller landowners rank enjoyment of forest ownership highest compared with larger landowners. Although it is very difficult to generalize, it appears that as tract size and value and frequency of timber

Table 1-4.
Ownership expectations of future timber harvest

Expected Future Harvest	Percent of Owners	Percent of Acres
1–10 years	31.6	63.1
Indefinite	27.7	23.1
Never	34.9	11.5
No answer	5.8	2.3

revenue increase, timber production becomes a primary reason for ownership.[1] Nonetheless, most of those owners also have multiple goals, combining timber production and other values.

Some interesting variations show up when the primary reasons for ownership are compared with owners' statements of the benefits they desire to derive from that ownership in the next ten years. Questioned in this way, more owners indicate their intent to gain income from timber harvest, increasing the acreage oriented to timber production from 29% to 33%. Strikingly, while only 9% of owners state that land investment is their primary reason for owning forestland, 20% expect to reap the benefit of increased land value in the next decade. This 20% appears to be weighted heavily toward smaller landowners. All in all, many owners are expecting greater productive uses of their forests in the coming period, increasing to 45% of ownerships and 63% of forest acreage.

It is also worth noting that expected enjoyment of forest ownership markedly increased as a primary reason for ownership, from 14% of owners to 34% and from 7% of acreage to 16%. This increase is probably attributable to the responses of many whose primary reason for owning land in the first place was incidental to ownership of a residence or farm.

1. In a study of 1,300 Virginia private forest landowners, researchers found that both harvesters and nonharvesters rated preservation of nature, scenic values, and wildlife as their top reasons for owning forestland (Hodge and Southard 1992). Similarly, a 1994 survey of primarily larger NIPFs in Indiana, Utah, and the Southeast found that the highest-rated reasons for forest ownership were preservation of wildlife habitat, maintenance of natural beauty, personal recreation, and simple satisfaction of ownership (Brunson et al. 1996).

Figure 1-5.
Percentage of ownerships and acreages with intent to
harvest within next 10 years by size class

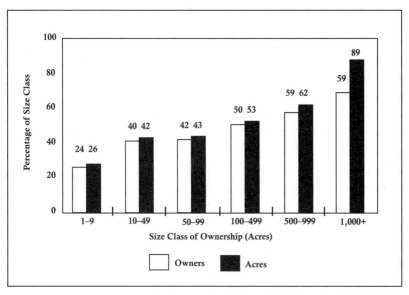

Timber Harvest Activities of Various Landowners

Looking at these data should be reassuring to those concerned about future timber supplies, as the vast majority of owners indicate their willingness to cut timber at some point. According to Birch (1996), 46% of forest owners, who own 78% of all forests, had previously harvested timber on their land. Table 1-4 shows that almost 60% of owners, with 86% of forest acreage, intend to harvest in the future. At the time of Birch's survey, only 11% of private forests were owned by people with no intent to ever harvest.

Figure 1-5 shows ownership organized by size class for both number of owners and acres in each class. We will consider timber harvesting further when we focus on the behavior and attitudes of individual nonindustrial owners in the next section.

In general, the harvest behavior of industrial and nonindustrial landowners is different. Industrial forestlands are owned primarily for fiber output to supply processing facilities; therefore fiber output is maximized to the degree possible. Nonindustrial forestlands are held for a wide variety of reasons. Ownership surveys find that in general NIPFs are not opposed to timber harvest. In the South, for instance, historical rates of harvest for industry and NIPF owners are comparable (Alig et al. 1990a). However, current research indicates that NIPFs value their standing timber more than industrial owners. This seems to be due to the value NIPF owners put on nontimber forest resources, "receiving non market (non measured) benefits from holding timber in place" (Newman and Wear 1993). In other words, while NIPF owners will harvest timber, they also highly value the amenities provided by the forest itself.

Degree of Forest Management Planning by Landowners

The USDA Forest Service estimates that in 1993 5% of owners had written management plans for their forests. These owners—most likely from the same group that gave timber production as a primary reason for and benefit of forestland ownership—hold 39% of private forests. Forty-three percent of them are industrial owners and 57% nonindustrial. Most forest management plans focus on timber harvest. It is not known to what degree ecological resources are included. Given that so many NIPF owners have multiple goals for their forests, with timber harvest included but not primary, there are great opportunities to expand owner engagement in forest management planning if a greater emphasis is placed on overall forest stewardship than on commercial timber harvest.

Length of Forest Ownership

Greater ownership turnover tends to lead to reductions in parcel size and increased fragmentation of forestland. Forests are turning over faster than it takes for them to mature. As each new owner takes title, new goals for the land are set. Inconsistent forest management and even overharvesting over time can be the result. The dates forest owners of all types acquired their forestland show that more than 40% of owners acquired forestland for the first time since 1978. These recent acquisitions involved 23% of private forest acreage (Birch 1996; figures 1-6 and 1-7). Only 30% of forest acreage has been held forty-five years or more, in less than 10% of ownerships.[2] We will look at some of the trends apparent in recent turnover at the end of this section.

---

2. Some 11% of private forests or 44.7 million acres has been held in the same ownership since before 1900 by an estimated 66,600 owners. Eighty-eight percent of these owners are farmers or individuals; 0.3% or two hundred are from the forest industry.

Figure 1-6.
Private forest ownerships by date of acquisition
(percent of owners in 1978 and 1994)

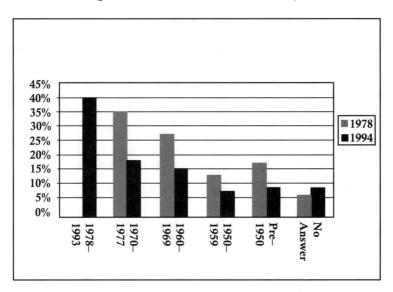

Figure 1-7.
Private forest acreage by date of acquisition
(percent of acres in 1978 and 1994)

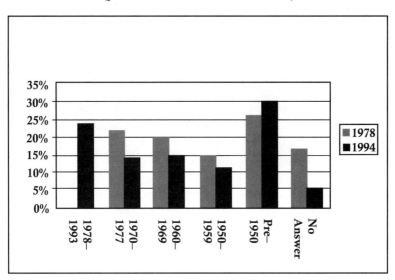

When the diversity of private forest landowners and the diversity of their goals in forest ownership are considered, it becomes easier to understand the impact of changing ownerships on the forest itself. Aside from the clear trend of more ownerships and smaller parcel sizes at every size class, turnover in forest ownership has other impacts. Whether large or small, industrial or nonindustrial, the land use decisions of each owner are imprinted on the forest and lasting in nature. The rate and intensity of timber harvest, road and home building, agricultural conversion, introduction of exotic species, and other activities often are compounded through time by turnover in ownership.

Focus on Nonindustrial Private Forest Owners

As of 1997, nonindustrial private owners held about 326.8 million acres of forestland, of which 290.8 million were classified as timberland. This represents 58% of all timberland. (See appendix table B-1 for state and regional details of forest- and timberland ownerships by acres.) Some 72% of the nation's hardwood inventory and 30% of softwood are found on nonindustrial timberland. About 60% of the commercial timber stocking on NIPF land is hardwood. Statistics do not yet capture the stocks of noncommercial species or species occurring on "other forestland."

Nonindustrial ownerships are most numerous in the East. About 42% of the NIPF forests are in the southern regions (comprising 70% of all forestland in the South) while 32% are in the North Central and Northeast regions (comprising about 67% of forestland in these states). Viewed another way, southern NIPFs own 49% of U.S. timberland; in the North Central and Northeast regions, they own about 40%. Western NIPF ownerships are 25% of total NIPF forests, controlling about 25% of forestland in their states.

Of the nontimberland held by private owners in the United States, 36% is in Alaska, largely held by Alaskan Native corporations. Another 25% is found in the Four Corners region (Arizona, Colorado, Utah, and New Mexico), largely as pinyon-juniper woodlands; 15% is in California's woodlands; and 10% occurs in Texas, largely as mesquite woodlands. For most of the remainder of the nation, NIPF forests are almost entirely classified as timberland.

As already noted, nonindustrial forestland owners are especially diverse. This discussion focuses most of its attention on the vast majority of NIPF owners, including residential forest owners who are individuals or families.

Table 1-5.
Size distribution of individual ownerships (1994)

Size class	Individual Ownerships	% Total Owners	Acreage	% Total Acres
1–9 acres	5,583,100	59.91%	15,847,000	7%
10–99 acres	3,212,500	34.47%	98,701,000	42%
100–499 acres	479,300	5.14%	77,137,000	33%
500–999 acres	28,900	0.31%	17,015,000	7%
1,000–4,999 acres	12,300	0.13%	17,051,000	7%
5,000+ acres	3,300	0.04%	6,596,000	3%
Total	**9,319,400**	**100%**	**232,347,000**	**100%**

Separately below we will examine two other important NIPF owner types: institutional investors and Native Americans. In understanding individual nonindustrial forest landowners, perhaps the most important thing to grasp is that they are essentially no different in their attitudes and sociodemographic profiles from Americans in general. There are some important distinctions, however. Generalizing the characteristics of some 9 million individual Americans who own forests obviously requires oversimplification, but the available evidence suggests the conclusions below.

America's forest owners are, like the general population, aging (figure 1-8). In 1994, 24% of the NIPF forestland was held by individuals over 65 years of age. This was up from 19% in 1978. Between 1978 and 1994 (figure 1-8) the amount of forestland owned by retirees increased from 47 million acres to 77 million.

Also, like the rest of the population forest landowners are more urban oriented (figure 1-9) than they used to be. Between 1978 and 1994, the amount of forestland owned by farmers and blue-collar workers dropped from 90 million acres to 60 million, while the amount owned by white-collar workers increased from 49 million acres to 68 million. It is likely that the increase in retirees has been drawn from rural, farming, and/or blue-collar owners as well as from new owners retiring from the city.

The average size of individual ownerships[3] has been shrinking for years and is now, or soon will be, under 20 acres (figure 1-10).

---

3. Including residential owners but apart from nonindustrial companies, private organizations, Native American tribes, and the like.

Figure 1-8.

Distribution of acreage owned by individuals by owner age class

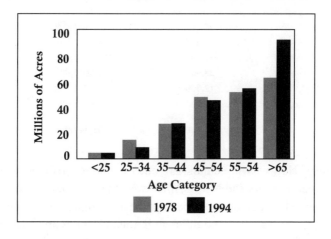

Figure 1-9.

Occupational categories of U.S. forest landowners by millions of acres owned

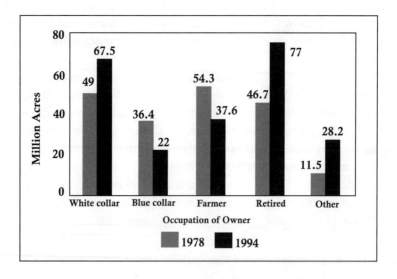

Figure 1-10.

Size of individual nonindustrial forest ownerships since 1953

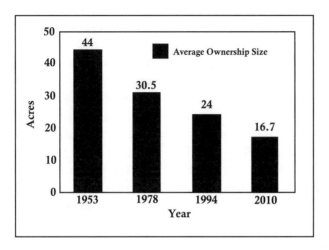

The environmental attitudes of NIPF forest owners are indistinguishable from those of the general population. In a conflict between environmental and economic interests, a majority of NIPF landowners think environmental interests should prevail.[4] (Bliss et al. 1997)

Larger forest landowners and those with a greater financial stake in timber harvest oppose government regulation as a means of achieving environmental goals (Bliss et al. 1997; Johnson et al. 1997). As the attitudes of NIPF owners toward the environment and regulatory protection of nontimber resources are important to understand, we will discuss them further below.

Contributions of Nonindustrial Owners to Timber Harvest

These landowners as a class consistently provided 47 to 52% of the timber harvested in the United States for the forty years from 1950 to 1990 (Alig et al. 1990a). As discussed further in chapter 2, during the 1990s, the NIPF share of harvests rose to 60%, a dramatic increase from historic levels driven by reductions in supplies from federal and industrial sources. This increased share is expected to continue for at least the next fifty years. Softwood sup-

---

4. In a 1993 Pennsylvania survey, forest landowners more than the general public were found to engage in environmentally prompted actions such as utilizing environmental criteria in their buying decisions. As one reviewer commented, "The 'typical' Pennsylvania landowner is apparently an environmental 'activist'" (Jones et al. 1995).

plies from NIPF ownerships have declined and hardwood supplies have increased. NIPF lands, which have 70% of hardwood stocks, are providing 75% of harvests. Because NIPF lands contribute such a large percentage of the nation's timber supply, the sustainability of NIPF lands is directly linked to the sustainability of the timber supply in the United States.[5]

More than 80% of these harvests occurred on larger ownerships where harvests may be more regular than the episodic harvests of small ownerships. In general, NIPF owners are more likely to harvest when the current market price for timber is high or if perceived threats to tree mortality by insects, disease, or fire are increased. Since timber harvest is not a high priority for most NIPF owners, they may tend to wait out market fluctuations hoping for a high enough price to ensure an economic return on their investments. Timber sales may be dominated as much by family or financial conditions (an owner's death or a major financial need) as by a forest management plan. However, there are few hard data on the relative economic importance of timber harvests on NIPF lands to the employment and income of the owners (NRC 1998).

Interestingly, other nonforest income may be a factor in harvesting decisions. In analyses of NIPFs in the East, higher-income owners appear less likely to harvest timber than lower-income owners (Alig et al. 1990). Higher education also appears to be negatively correlated with timber harvesting by individual forest owners (Binkley 1981; Boyd 1984). At the same time, landowners with higher income and/or higher education demonstrate a greater willingness to learn and innovate. Therefore, their interest in timber harvest tends to be tied to learning about their forest property and gaining technical assistance in forest management (Hodge, pers. comm.).

Individuals with small- to medium-sized acreages appear more likely to engage in active forest management when they understand its role in the context of their forest stewardship goals and not simply as logging for financial remuneration (Bourke and Luloff 1994). Frequently the timber

---

5. It is worth noting that almost all the many studies of nonindustrial forest landowners are oriented to understanding why they do or do not cut timber. As the authors of one study in Pennsylvania noted in the *Journal of Forestry,* "These studies have often been framed in the context of solving the NIPF 'problem'—generally referring to the role of NIPFs in contributing what is perceived to be their share of wood products to society. Until recently such investigations have rarely been motivated by sociological interests, the objectives of the private landowner, or the noncommodity and ecological values of their forests" (Egan and Jones 1993).

harvest goals for these owners reflect their multiple goals for their property: they desire some revenue, but they also want to enhance amenity, wildlife, and recreational values. Perhaps also reflecting the precedence of environmental and amenity goals, most NIPFs—even in the South—dislike or even oppose clear-cutting (Jones et al. 1995). However, even though timber harvest is not their main purpose in owning forestland—and they may have different silvicultural preferences than industrial owners—clearly NIPF owners do not oppose harvesting per se and intend to harvest at some point.

Attitudes of NIPF Owners toward Environmental Protection and Environmental Regulation

In reviewing the literature of NIPF research, it becomes clear that owning and managing forestland does not strongly influence attitudes toward forest management and forest policies. However, NIPFs appear to be slightly more conservative than the public at large (Bourke and Luloff 1994). While NIPF support for environmental protection is generally strong, using regulation as the means to that end is not supported by those owning larger forest properties.

As summarized by Bliss, NIPFs "share the public's concerns about clearcutting and herbicide use, support regulating forest harvesting practices where necessary on private land to protect environmental values, and generally value environmental protection highly relative to both private property rights and economic development" (1997). The data suggest this is true regardless of gender, income, or residence.

In Bliss's Tennessee Valley study of NIPFs, strong majorities agreed that private property rights, while important, were secondary to environmental protection and that rights should be limited where necessary for the environment. However, as tract size increased, and with it timber orientation, property rights sentiment increased. When the sample was limited to landowners with 100 acres or more, only 27% supported regulations as a means to protect water quality, threatened species, and scenic beauty.

These findings were corroborated by Johnson et al.'s survey of nonindustrial landowners in the Pacific Northwest (1997). They found that the strong majority of landowners were not influenced by the threat of future regulation in their recent harvest decisions. Yet as their financial stake in the forestland increased (larger tracts, majority of income from timber, long-term hold, mature trees), they became more concerned about the impact of possible future regulatory restrictions on their investments. About 75% of the larger landowners (those with more than 100 acres) felt that no additional restrictions should be put on private lands to protect riparian ecosystems or endangered species. A majority indicated they would harvest sooner than planned if they felt such new regulations were imminent. Nonetheless, 58% of larger and 70% of small landowners agreed with the statement, "I would be willing to alter the amount and time of my harvest if it is necessary to maintain a healthy ecosystem."

These insights into NIPF characteristics will be especially important as we consider the trends in forest ownership and their implications for a conservation strategy.

The Nature and Goals of Institutional Forestland Owners

Institutions such as pension funds, foundations, university endowments, and the like are a small but growing class of large, nonindustrial landowners that control an estimated 5 to 7 million of the 352.5 million acres of NIPF forests. Since the passage of the federal Employee Retirement Income Security Act (ERISA) in 1974, pension fund ownership of forestland has grown as an investment asset included for diversification within these owners' huge portfolios. When the Hancock Natural Resource Group first organized a timber investment fund for large institutional investors in the mid-1980s, total institutional investment in forests was an estimated $300 million. It grew to an estimated $7 billion in 1999. This represents some 1% of pension fund assets and a similar percentage of the estimated total U.S. private forestland market value. CalPERS, the huge California public employee pension fund, and Ohio State Teachers Fund are among the largest of this class of owners, acting through their timber investment managers. A handful of managers represent institutions. In addition to Hancock, the major ones include UBS Brinson, Wachovia, Prudential, the Forestland Group, Forest Investment Associates, Wagner Woodlands, and the Campbell Group. Collectively, these managers are called timber investment management organizations (TIMOs).

Institutional owners have been drawn to forest ownership because of the perceived characteristics of forestland as an asset class within their portfolios. Economists have analyzed the behavior of forestland compared with other financial assets and found that it can provide relatively high risk-adjusted returns, especially with holdings that diversify commercial species and regions. Pension fund investors, by far the largest institutional forest landowners, are in a fiduciary role, representing the many beneficiaries of their institution. They generally take a longer-term view of their investments and have historically been very averse to risk. To generate the highest possible returns while mitigating risk, institutions have developed very sophisticated financial models to guide their acquisition and disposition of a wide range of assets.

The most basic goal of institutional forest landowners is to deliver a desired rate of return from the sale of timber and land while minimizing risk. They do not have mills or other processing facilities to supply. Therefore, although among the largest landowners, they have relative flexibility in merchandising their timber and land. Like smaller NIPF owners, but with much greater scale and sophistication, they can choose to sell or not sell commercial tree species and different kinds of timber for particular products as they read the markets. Similarly, they will move in or out of a specific forest type, region, or, as is now happening, country to fit their financial models.

There have been no studies of these owners to better understand their actual forest management behavior or the importance of nonfinancial goals relative to financial ones in their decision making. In general, institutional owners have tended to manage forests on an industrial model, although there are exceptions. As fiduciaries, their ability to invest in activities that do not bring direct returns is limited. As managers for high-profile, quasi-public institutions—with the retirement funds of many individuals in their care—good government, community, and public relations are important for them to maintain. Therefore, institutional owners engage in forest stewardship activities to demonstrate good citizenship and mitigate risk at the same time. We will discuss the growing institutional ownership of forestland further as we consider overall trends in forest ownership below.

Focus on Industrial Private Forest Owners

About 67.6 million acres or almost 9% of U.S. forestland are owned by the timber industry, of which 99% is timberland (USDA, Forest Service 2000). This represents 13% of all timberland. (Birch estimated almost 80 million acres in 1994 using a somewhat different definition.) Between 1952 and 1992, the forest industry acquired 11.5 million acres of private forestland from larger NIPFs (whose ownership acreage decreased by 16.8

million). Industrial ownership is greatest in those regions with highly valued commercial timber species and highly productive soils. In the Pacific Northwest, South Central, and Southeast regions, industrial owners control 18 to 19% of all forests. Industrial owners are especially significant in Maine, where they own 46% of forest area. Industrial ownerships include almost 14% of U.S. softwood inventory and 9.5% of hardwoods. About 70% of the commercial stocking on industrial lands is softwood. During the 1990s, forest industry lands produced one-third of the total timber in the United States.

U.S. forest products companies are among the world leaders, both in the highly competitive pulp and paper business and in the timber-products business.[6] This section looks at the attributes of larger industrial forest landowners, from publicly traded transnational companies to regional private companies.

As with other ownership types, the nature of the industrial forest is dictated by the management objectives of its owners. These objectives are somewhat more easily quantified than with nonindustrial owners. By definition, the primary goal of industrial forest ownership is production of wood and fiber to supply owners' mills. Therefore, forest management is geared to increasing yields per acre. This is accomplished through increasingly intensive management, as well as by buying and selling forestland to maximize the ownership of highly productive and easily operable lands—in the process divesting lands less suited to high-yielding timber production. Although they do not own mills, the largest nonindustrial owners generally engage in similar practices.

The forest management practiced on industrial lands is influenced greatly by the processing facilities owned by each company. Within this category of forest ownership there are two basic types: sawtimber producers and pulp producers. Although many paper-oriented companies have both divisions, the relative weighting within a company can often affect its forest management goals. Simply put, sawtimber companies tend to manage for desired dimensions of trees so as to increase yields of par-

6. Already by 1991, according to *Pulp and Paper International,* U.S.-based companies accounted for seven of the top ten pulp and paper companies with international holdings. These included International Paper, Champion (now combined), and Georgia-Pacific. As of the late 1990s, five U.S. companies dominate the world sawmilling capacity as well: Weyerhaeuser, Georgia-Pacific, International Paper, Sierra Pacific, and Louisiana-Pacific (Crossely and Points 1998).

Figure 1-11.

Average timber growth on industrial lands in three regions (1952–1992) (cubic feet/acre per year for all species)

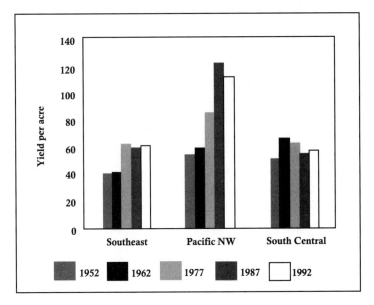

ticular lumber products, while pulp and paper companies tend to manage for cubic footage of fiber per acre.

The forest products industry is very capital intensive. Manufacturing facilities and forestland both require hundreds of millions of dollars to build or acquire. The pressure to provide a return on such substantial capital investment is enormous. Bricks and mortar, as well as biological resources, are not very liquid; therefore the business challenge of building cash flow is preeminent.

The age of a forest and the sizes of its trees are determined through the owner's manufacturing needs and desired rate of return. In general, an industrial forest that has been produced is younger, with smaller trees and a simpler species composition relative to its biological capacity. Correspondingly, industry has reengineered to focus wood production on a new suite of products that can utilize low-quality wood or fiber in ways that mimic the characteristics of now-scarce high-quality, older trees. Engineered wood products, such as laminated beams, medium-density fiberboard, and oriented strandboard, are rapidly supplanting plywood and larger-dimension lumber. Many companies that formerly managed for sawtimber have therefore shifted their management orientation to fiber.[7] With quality built at the factory and not in the forest, increased utilization of formerly less commercial tree species is possible. Engineered wood products also diminish processing waste, allowing industry to complete utilization of harvested trees.

Faster growth rates mean better fiber flows and improved financial results for shareholders. Investments by the forest industry in improving tree growth include species selection, plant breeding, and genetic engineering to produce significantly faster-growing trees. For instance, plantations of genetically altered loblolly pine can produce 26% greater yields than those grown from wild seed (Guldin and Wigley 1998). The forest products industry also uses other intensive management practices such as matching species and genotypes to specific soils during tree planting, controlling competing vegetation, timely thinning to keep stands free to grow, and fertilization. The effects of these practices can be seen in figure 1-11, which illustrates the trends in average annual timber growth of all species on the industrial lands in three regions since 1952. Note that, while plantation forestry featuring loblolly pine has been gradually changing southern growth rates, the conversion of older natural forests to younger, and therefore faster-growing, Douglas-fir plantations along with the introduction of "super trees" has had a dramatic effect on average annual timber growth in the Pacific Northwest. In 1992, timber growth in the Pacific Northwest was twice the national industry average, and five times the national average. We will discuss timber productivity and the sustainability of supplies in chapter 2.

Plantation forestry is increasingly preferred by the forest industry and other large commercial forest ownerships. Over the last twenty years, remaining older natural forests are being converted to short rotation management to increase fiber flows and improve financial returns by tying up capital on the ground for shorter periods. Management

---

7. Interestingly, after a trend toward reducing harvest age, pulp and paper companies in the Southeast are beginning to expand their rotations from twenty to thirty years in order to build in flexibility to capture the value of both fiber and small-dimension sawtimber utilization (C. Owen, pers. comm. 1999). Meanwhile, some sawtimber companies in the Pacific Northwest have been reducing notation ages from fifty-five to sixty years to forty and now thirty years.

inputs have been intensified as described above. According to recent USDA Forest Service data, about 36% of forest industry softwood acreage was in plantation, accounting for 21% of volume. This compares with 12% of NIPF acreage and 7% of inventory (Haynes 2001). Plantations account for almost all industrial (and large nonindustrial) Douglas-fir forests under fifty years old in the Pacific Northwest. The USDA Forest Service projects that the total area of industrial plantations will increase by 76% and the share of inventory will triple (Haynes 2001).

Some information about the management activities of industrial private forest owners is available through the Sustainable Forestry Initiative Program (SFI), operated by the American Forest & Paper Association (AF&PA), the largest trade association for the industry. Information is provided by companies on activities such as the extent of different harvesting practices, reforestation, and compliance with water quality best management practices (BMPs). As of 1998, some 120 companies were listed as SFI program participants, representing about 56 million acres of forestland (AF&PA 1999). On those lands, SFI participants reported harvesting about 1.3 million acres in 1998. These harvest acres include all types of harvest, including partial thinning and salvage. Reforestation (both natural and planted) was carried out on about the same amount of land. SFI program standards call for fully restocked reforestation by appropriate means within two years, and the industry is able to realize that target more than 98% of the time, according to the SFI progress reports.

Some industrial forest landowners have developed recreational revenue sources from their lands as a complement to timber production. This is true for other large forest landowners as well. There is also increasing development of special forest products revenues, especially from the production of items for the floral industry. In certain regions, such as the Great Lakes and New England, public access for recreation is either a tradition or a requirement of preferential forest property tax treatment.

Although the financial incentives are compelling for industrial forest landowners to intensify forest management and simplify the species composition and structure of forests to maximize commercial outputs, active stewardship of noncommodity forest values is on the rise. Government regulation of private forests has increased dramatically in the

Table 1-6.

Forms of Native American forest title

Type of Ownership	Acreage	Title Status
Tribal trust	14,488,000	U.S. government
Individual trust	865,000	U.S. government
Individual restricted fee title	868,000	Native American with U.S. government restrictions
Tribal restricted fee title	6,000	Tribe with U.S. government restrictions
Tribal fee simple	820,000	Tribe without restrictions

last twenty-five years, since the passage of the Endangered Species Act (ESA), Clean Water Act, and other state environmental legislation. The public has expressed its desire to sustain wood and fiber production within the context of ecological sustainability. Operating at or above the stewardship standards set by law has become paramount for industrial operations to maintain their "social license to operate." Building positive regulatory and community relations is essential to mitigating risk and maintaining consistent production in their operations. Most industrial forest companies understand that private property rights come with responsibilities.

With the historic legacy of public mistrust, industrial forest products companies are more positively engaged in activities to protect or restore habitat and water quality than ever before. AF&PA reports that in 1998 12.3 million acres of land in its SFI program were covered by some kind of cooperative fish and wildlife management agreement with a government agency or conservation organization. The forest products industry has a clear preference for voluntary approaches to protecting and enhancing environmental values on its lands.

Similarly, over the last eight years the forest products industry has increasingly utilized Habitat Conservation Plans (HCPs) to meet the requirements of the ESA while maintaining its forest management. According to a review of HCPs performed by PFT in October 1999, approximately thirty-one incidental take permits on forestland had been approved under Section 10 of the ESA, covering management of more than 8.5 million acres. Much of this is industrial forestland, including property owned by International Paper, Potlatch, Weyerhaeuser, Simpson, Union Camp, Crown Pacific, Plum Creek, and Pacific Lumber.

In terms of land use dynamics, industrial forestlands are generally considered more stable than many other private lands as they are usually owned in large tracts for the express purpose of producing long-term timber supplies. Most acquisitions and dispositions are between industrial owners, or between industrial owners and large nonindustrials. That does not mean, however, that industrial lands are insulated from significant change. Corporate takeovers, mergers, and land sales occur regularly. The increasing rate of turnover in this ownership class is discussed at the end of this section.

In the course of these transactions, sellers typically analyze what portions of their property may have highest and best use development value other than as timber. Such sales can be highly profitable. While it is difficult to track, anecdotal data gathered by PFT suggest that some 5 to 15% of divested lands were sold as real estate. These conversion transactions tend to occur in the expanding edge of the urban-wildlands interface and in regions where forest properties have high recreational value (especially along rivers and lakes, and in coastal areas).

Focus on Tribal Forest Owners

It has been estimated that in the continental United States, 193 Native American reservations[8] in twenty-three states contain some 16.8 million acres of forestland. Of these, 5.7 million acres are managed for timber production, 1.7 million acres are noncommercial timberland, 4.4 million acres are commercial woodland, and 4.2 million acres are noncommercial woodland (Morishima 1997; Intertribal Timber Council 1993). Native American ownership takes the five forms shown in table 1-6.

Ownership fragmentation is a major problem on the forestland owned by individual Native Americans, often held in trust by the federal government. These lands were distributed to native individuals in small allotments (often 160 acres) under the 1887 Dawes Act. As each generation passes, and those allotments are divided up among heirs as undivided property interests, administration of federal trust responsibilities becomes more of an administrative nightmare. The U.S. Department of the Inte-

---

8. *Forested Landscapes in Perspective* (NRC 1998) puts the number of reservations at 214 in 1992.

rior, charged with collecting fees and redistributing them to the appropriate owners, has been unable to carry out that responsibility for decades, in spite of legal action demanding that the situation be rectified. It is uncommon for these fragmented ownerships to be divided up and sold for development, as is often the case with non–Native American forests. Such ownership fragmentation makes decision making and management increasingly difficult.

Tribal leaders have opposed the adoption of certain aspects of private property principles that are typical of non-Native American U.S. ownerships. While it is difficult to generalize across a wide diversity of native peoples, tribes typically do not want tribal members to sell land without permission of the tribal government. They also prefer that the lands held by Native Americans be held by the tribe and not by individuals. Both preferences are rooted in cultural traditions (NRC 1998).

As sovereign nations, tribal reservations govern themselves and are not subject to land use or forestry regulation by the states. They are, however, subject to relevant federal laws, either directly or through the Bureau of Indian Affairs (BIA). Given the history of European settlement and conflicts with Native Americans, tribes are very concerned with establishing and maintaining control of their land and resources. It is important to remember that treaty terms are still being litigated by tribes for enforcement of their rights. Fishing and hunting rights based on customary tribal use have been upheld in the courts and through statute in several states. There are many controversial issues on the relationship of tribes to federal and state governments, as well as to surrounding nontribal communities.

As with other forest landowners, tribes have various goals for their forests, both cultural and economic. While Native American timber makes only a modest contribution to the national totals, it is an extremely important source of revenue for the tribes, generating some $465 million and 40,000 jobs for tribal communities in 1991 (Morishima 1997). In addition, many other commercial and noncommercial uses of Native American forests and woodlands are important contributors to both cultural and subsistence needs. Fish, wildlife, medicine, native foods, and firewood are of prime importance to most Native American communities.

Forest management on Native American lands was, for many years, carried out almost entirely by the BIA, which was subject to serious and consistent underfunding from Congress. A 1993 study by independent forestry experts concluded that Native American forests were receiving 37% less funding for timber production than national forests, and only 50% of what was being invested on private lands (IFMAT 1993). That report concluded that the federal government should turn primary responsibility for management of Native American lands over to tribal forestry programs, a move encouraged by federal laws and supported by the Intertribal Timber Council (ITC), a consortium of seventy-three Native American tribes and Alaska Native organizations (Morishima 1997). According to a 1997 assessment by the original IFMAT team (Gordon et al. 1997), however, there is still much to be done in implementing the report.

Tribal-based forestry programs for Native American lands have evolved considerably over the last decade as tribes assert more control over their resources. Nonetheless, even within tribes there are conflicts among leaders with economic development goals and those with cultural or ecological goals. Some Native American tribes have a long tradition of commercial forest management that includes strong ecosystem values. For instance, the Menominee of Wisconsin and the Yakima of Washington have been cited for their outstanding forest management programs, demonstrating in many cases that production of vital timber supplies does not preclude excellent ecosystem management, species protection, and protection of cultural values. The Menominee Tribal Enterprises and the Hoopa Tribe of California have, for example, been certified under the Forest Stewardship Council's (FSC) program. They report that forest certification has opened up markets for new secondary products, increased the value of some less marketable species, and led to more forest jobs and products.

Trends in Forestland Ownership

Several large undercurrents in U.S. private forest ownership deserve to be highlighted. Just as forest ownerships, both public and private, form an interconnected mosaic, these trends interact as well. Their cumulative effects compound each individual trend. We will examine in turn the growing fragmentation, sprawling development, and de facto conversion of forest ownerships; the aging of individual forest owners; the restructuring of the forest industry; and the rise of financial owners. We will also note the emergence of conservation ownerships. The implications of these

Table 1-7.

Comparison of private forestland acres and ownerships in the United States, 1978 and 1994, by size of ownership (acreage category)

Acreage Category	1978 Owners	1994 Owners	1978 to 1994 Change	Percent Change	1978 Acres	1994 Acres	1978 to 1994 Change	Percent Change
1–9	5,765,000	5,686,000	(79,000)	(1.4%)	11,457,000	16,022,000	4,565,000	39.8%
10–49	1,213,000	2,644,800	1,431,800	118.0%	29,164,000	57,848,000	28,684,000	98.4%
50–99	477,000	685,000	208,000	43.6%	34,371,000	45,296,000	10,925,000	31.8%
100–499	566,000	527,000	(38,600)	(6.8%)	106,933,000	87,635,000	(19,298,000)	(18.0%)
500–999	40,000	37,800	(2,200)	(5.5%)	28,122,000	22,952,000	(5,170,000)	(18.4%)
1,000+	24,000	23,600	(400)	(1.7%)	144,139,000	139,213,000	(4,926,000)	(3.4%)
Total	**8,085,000**	**9,604,600**	**1,519,600**	**18.8%**	**354,186,000**	**368,966,000**	**14,780,000**	**4.2%**

trends will be discussed in the context of threats to America's forest resources in chapter 3.

Forest Ownerships Are Fragmenting

According to an analysis prepared for this book by Thomas Birch, the lead ownership researcher for the USDA Forest Service, and summarized in table 1-7, total numbers of forest owners increased by an estimated 15% in the fifteen years between the 1978 and 1994 USDA Forest Service studies. During this period there were dramatic increases in ownerships and forestland held by the smallest owners (those with less than 50 acres): approximately 116% more owners and 138% more forestland. In general, there has been a downsizing of all ownership classes of 100 acres and larger. Although the medium forest owners (those with 100–999 acres) were most affected, with 24.5 million acres being lost to smaller parcels, even the largest landowners were reduced in numbers and extent.

To create table 1-7, Thomas Birch worked with us to analyze and provide a better statistical match between the data from the 1978 and 1994 studies of private forestland owners. Before this effort, a simple comparison of ownership data for the two studies showed an increase of some 60 million acres of forestland that is merely an artifact of two different forest definitions used in the studies. Although even the new analysis shows some increase in the total forestland base, it is minimal. We believe the comparison used in table 1-7 more accurately describes the changes in private forest ownership that can be deduced from available data.

To accomplish more of an "apples-to-apples" comparison between the studies, the apparent increase of 60 million acres of forestland in 1994 than in 1978 had to be addressed. Although 10 million acres of this total were due to changes in the classification of Native American lands, most of the difference lay in how "other forests" were treated in the two surveys. Further complicating a comparison, in 1978 the study used the estimate of forest landowners developed by the NRCS's National Resources Inventory, while the 1994 base dataset came from the USDA Forest Service's Resource Planning Act Assessment. In table 1-7, Birch used the RPA landowner numbers in both years, added Native American lands, and limited the inclusion of nontimber forestlands from the 1994 numbers. Native American forests increased by 7 million acres in 1978 and 17 million acres in 1994, almost entirely in the 1,000 acre-plus category. This was due to the (continuing) disposition of Alaskan forestland as part of the Native Claims Settlement Act.

Unfortunately, it is very difficult to completely reconcile the datasets, and "other forests" east of the Great Plains states were still included in the 1994 data. The inclusion of the "new" Native American forests and the eastern "other forests" gives the appearance of increasing forestland by some 14.7 million acres or 4% over 1978. The new Native American forests add ownership and acreage at the larger size categories, while eastern nontimber forests tend to increase ownerships and acreages at the smaller. These statistics illustrate the continuing shift to smaller, more residential ownerships and the concurrent reduction in the midsized, mostly nonindustrial ownerships. There are growing numbers of individual owners. While there is no clearly georeferenced data, it is evident from regional analyses that these parcels are growing along the urban-rural interface and in the accessible rural areas, especially along rivers and lakes.

Given the characteristics of individual forest landowners, it is likely that today more forestland is owned by individuals and families who are less interested in timber harvesting as an important revenue source than was the case in the past. Many of the new forestland owners are likely to be more urban in orientation than owners of the past, given the increasingly urbanized nature of our society and the growing white-collar occupations of forest owners. These new owners may also be more concerned with the ecological and amenity values of their forests than were previous owners, given the rough correlation between size of ownership and primacy of timber-related goals.

Regardless of the characteristics of the many new ownerships, smaller parcels and varying landowner goals in and of themselves create new issues for forest conservation. With more small parcels comes a denser and more extensive patchwork of built infrastructure and other nonforest features that alters larger-scale forest ecosystem functions. These include impacts to hydrological functions, fire regimes, and habitats for interior and wider-ranging native species. At a certain point the impact is great enough to consider the parcel to be converted from forest.

The sprawling nature of most urban development in the United States exacerbates the reach of the impacts, pushes the leading edge of nonforest development, and accelerates fragmentation of larger, intact properties. Large-scale fragmentation limits the ability of interested individual landowners to realize certain environmental goals within the patchwork of small parcels.

Individual Landowners Are Aging

As illustrated in figure 1-8 earlier in this chapter, as of 1994 an estimated 2.5 million individual forest owners were 65 years and older (27% of all individuals) and held 92.6 million acres or 23.5% of the total privately owned forests. At that time, another 2 million owners were estimated to be 55 to 64 years old, controlling an additional 54 million acres. Virtually all of the former and some of the latter properties will go through some sort of intergenerational transfer in the next twenty years. Any property transfer is a moment when tracts can be broken up and parcel sizes reduced. In the context of an estate settlement, this is all the more likely because of the host of succession issues: too many heirs, no heirs; heirs' competing interests in property, no interest in property; and/or insufficient nonforest funding for estate taxes.

The scope of impact of intergenerational succession issues will vary from family to family, depending on the size and value of the forestland and its relative value within the estate. At current estate tax rates, and assuming a proportional representation of senior owners across size classes, our analysis suggests that unfunded estate taxes could force some degree of subdivision or unplanned timber harvest in the not too distant future of perhaps 5.5 million acres in the 1,000-acre-plus size class, owned by 4,000 people or families. This represents an estimated 1.5% of U.S. forestland. An additional 7,300 owners with 4 million acres in the 500- to 999-acre size class could also be affected, depending on timber stocking and value.[9]

Of the various age classes of individuals, Birch found aging forest owners to be the most likely to have harvested in the last decade and to have harvested the most acreage. Given that most of these people are retired, their harvest behavior is not unexpected as a source of supplemental income. A study of individual forest landowners in California by Romm et al. (1983) found that landowners over 65 years of age were also the least likely to make stewardship investments in their timber or other forest resources. Once more, however, it may be dangerous to extrapolate too broadly from the available data on private owners. A countertrend of increased stewardship activities by seniors might be under way if one concludes that retirees today include better educated, wealthier individuals who overall show a greater interest in active stewardship.

The Forest Industry Is Restructuring

Most of the decade of the 1990s was characterized by very poor performance of pulp and paper companies. The industry was barely covering its cost of capital and generating little if any free cash flow. Pulp prices were wildly volatile. Wall Street was not happy and bid down the stocks of publicly traded forest products companies, demanding improved returns to shareholders. Even the expanding economy was not solving these major structural problems. The result has been an acceleration in industry consolidation and turnover in U.S. industrial forestland, both part of company strategies to improve shareholder returns and compete successfully in an increasingly globalized business sector.

---

9. This is based on an estimated average value per acre of $2,000. Therefore, at current levels of estate tax and a unified credit of $1 million for a family business or for individuals (as is being phased in for individuals through 2006), single owners of more than 500 acres begin to incur tax liability, exclusive of residential value or other assets and assuming no estate plan is in place. The ability of landowners to pay the estate liability will, of course, depend on both the nature of their timber and nontimber assets.

A number of factors are at work here. As has historically been the case, once more forest products capital investment is migrating to regions with lower costs and cheaper resources. To compete with low-cost producers in Asia and South America, U.S. companies have been upgrading the efficiencies of their processing facilities. In the course of this process, many older mills have been closed. For instance, Louisiana-Pacific "shuttered half its 40 woods products mills during the past three years" (Starkman 1999). Companies also are seeking out the highest-yielding soils and species worldwide to grow fiber as fast as possible. U.S. forestland that is being retained is usually the highest site class and the most operable. These forests can be managed most intensively, yielding more fiber on less acreage. Others have found it cheaper to import pulp than to own U.S. forestland. This has fueled some of the dispositions of forests described below.

Historically the sector has been characterized by numerous local, regional, and international companies. Over time it has come to suffer from global overcapacity. Now larger companies have taken the course of industrial consolidation to gain efficiencies and greater control over the resource and markets. A wave of mergers has swept this sector in the last five years. Industry giant International Paper's acquisition of Union Camp and its purchase of prime rival Champion epitomizes the trend. Other notable recent combinations include Jefferson Smurfit and Stone Container's merger (Smurfit–Stone Container) and Weyerhaeuser's acquisition of MacMillan-Bloedel. "The big forest products companies are getting bigger. The middle-sized players are disappearing. The small players must be nimble, flexible, and entrepreneurial to compete effectively" (Collins 1999).

Timberland is carried at cost on the balance sheets of public forest products companies. Many of them have experienced great appreciation of these assets over the last twenty years, yet they have not been able to fully capture that value for their shareholders. To unlock the increased market value of their forests, many large industrial companies are divesting themselves of these lands or restructuring the ownership in such a way as to gain greater investor value. A series of huge transactions have been occurring and are expected to continue as companies sell off or otherwise monetize forests that are no longer strategically important to own directly. In the course of these dispositions, portions of the properties have gone into residential and recreational uses. At the same time, U.S. companies are expanding their holdings in Canada, South America, New Zealand, and Australia. (Meanwhile, Canadian companies are acquiring large tracts of the northern forest, notably J. D. Irving's purchase of the Sappi lands, making it the largest non-U.S. forest owner in the United States.)

When all kinds of dispositions are taken into account, an estimated 28% of industrial forestland—or some 20 million acres, an area four times the size of Massachusetts—changed hands in the 1990s. Approximately 18% of the state of Maine turned over in two years. Virtually all major forest products companies have sold major U.S. tracts in recent years. Pulp and paper companies have been net sellers. The majority of the lands have been acquired by other industrial owners, as well as institutional investors. For example, Georgia-Pacific spun off all its lands into a separate letter corporation, and then over three years sold its timberlands completely, with the California portion bought by a TIMO on behalf of a pension fund and the remainder acquired by Plum Creek. Louisiana-Pacific has also sold its California holdings.

Financial Ownership of Forestland Is on the Rise

Forestland has chiefly been a personal and industrial asset. It is now evolving into a financial asset, owned for its value as a portion of a diversified investment portfolio. Secondary markets are developing for units of forestland partnerships or similar entities. As discussed earlier, this is occurring for a variety of reasons. In addition to increased understanding of the role forestland can play in institutional portfolios, this ownership type is increasing as investors seek greater tax efficiency and liquidity in their holdings. By reducing the level of tax applicable to forestry profits, after-tax returns increase. Financial investors, with no ties to forest products processing, have great incentive to use various pass-through investment structures that are not subject to the double-tiered taxation to which investors in the conventional corporation are. Moreover, pension funds—the largest financial owners of forestland—are tax-exempt investors. Another financial advantage that some institutional ownerships derive from their form of organization is the ability to annually capture the increased appreciation in their land and timber through an appraisal process that allows them to include this unrealized appreciation in their calculation of internal rates of return. Corporations, especially publicly traded ones such as many pulp and paper companies, are unable to utilize this valuation method, being constrained by generally accepted accounting procedures to valuing their forest assets at their basis.

A variety of financial vehicles and products are available or evolving for those who want to make a pure play in forestland ownership, but in a way that is more diversified, with lower risks and greater potential liquidity than direct ownership allows. These include large private limited partnership funds organized by TIMOs; publicly traded master limited partnerships (MLPs), such as Crown Pacific; and, with the reorganization of Plum Creek, publicly traded timber real estate investment trusts (REITs).

With growing investments by pension funds and with the advent of more publicly traded, tax-efficient forest investment vehicles, it is likely that financial ownership of U.S. private forests will accelerate. Many of the forestlands from which industrial corporations have divested themselves in the South, Northeast, and coastal California have gone into financial ownerships.

Large-scale financial ownership does not in and of itself further forest conservation and sustainable management versus other forms of ownership. These owners are distinguished by not being tied to supplying a particular mill. They also have other management and marketing flexibilities that both large industrial and small nonindustrial owners generally lack.

Financial ownership has the potential to provide forest landowners with more consistent cash flows and investment liquidity than is possible through smaller, less diversified ownership or through other ownership structures. For smaller ownerships, liquidity is obtained through selling the timber or property. Sometimes the demand for liquidity leads to parcelization or overharvest. By assembling a large and diversified portfolio of forest holdings, cash flows can be smoothed while still harvesting timber sustainably. A secondary market for ownership units is also becoming available, preventing disruption of the forest due to change of ownership and thereby potentially promoting more consistent forest management.

On the other hand, more retail and indirect ownership of forests centralizes management control with financial managers, not with foresters, families, or communities. There is no guarantee that the demands of the capital markets for return from these forests will be any different than the demands experienced by the forest industry. Investment managers are evaluated by their ability to achieve certain benchmarks of return. Therefore, there is no reason these owners would have a longer-term perspective in their forest management than other large forest owners such as forest products companies.

Further, the market success of MLPs and REITs is driven by their distributions to unitholders. This puts pressure on forest managers to schedule timber harvests to meet the distribution objectives.[10] For publicly traded timber entities, markets will still be especially challenged to provide sufficient recognition of the value of standing timber and other forest assets versus timber harvest cash flows (Best and Jenkins 1999).

Conservation Ownership of Managed Forestlands Is Emerging

There is a small but noteworthy emergence of ownership of private, managed forestlands by conservation organizations. Traditionally, nonprofits have acquired environmentally valuable lands for transfer to public agencies. The Nature Conservancy and the Trust for Public Land are major examples of this kind of organization and transaction. Fee title lands held by conservation organizations have typically been managed as ecological preserves or quasi-public parklands. In the late 1990s, The Nature Conservancy, the Conservation Fund, and the Vermont Land Trust acquired private "working" forests in the Northeast. While portions of these acquisitions are going into public ownership, the remaining forestlands are either (1) being held and managed for sustainable forestry and conservation purposes as demonstration forests; or (2) being resold to other forest landowners subject to conservation easements that restrict subdivision and guide forest management to protect ecological values.

In three transactions, approximately 506,000 acres of forest were acquired. Although only a small percentage of recent forest transactions, these high-profile acquisitions may be creating a new model of forest ownership that combines high standards of forest resource protection with commercial forestry operations. See the appendices for more information on these transactions.

---

10. The experience of U.S. Timberlands, a publicly traded partnership founded in 1996, is a case in point. As of 1999, the partnership owned 673,000 acres of forestland in Oregon and Washington. Its unitholders had rights to distributions of $.50/unit per quarter. In order to maintain this level of cash flow, the company had to log at a much higher level than described in its offering prospectus. According to filings with the SEC, the company in fact logged at higher levels than its own estimated annual sustained yield of 110 million board feet. Its reported actual harvest levels have been 139 million board feet in 1997 and 145 million in 1998, with a planned 202 million board feet in 1999.

Chapter Two

The Nature of America's Private Forests

U.S. Forest Area, Regions, and Major Forest Types

The forest area of the United States is vast and varied. As with its owners, it is difficult to characterize simply without considerable generalization. However, some understanding of the nature and dynamics of our forest resources is essential to crafting the strategy for private forest conservation.

While unevenly distributed, forestland covers portions of all regions of the country. The arid interior West and the Great Plains are sparsely forested, while the more lush climates and rich soils of the Southeast and coastal Pacific Northwest generate dense forests. Although very different in character, all forests have ecological and economic value to our society. As regions, the Northeast and Southeast have the greatest proportions of forestland (table 2-1).

Forestland is typically classified by government agencies to reflect its status in relationship to timber production. Therefore, forests are categorized as either timberland, noncommercial woodlands, or areas reserved from timber production for use as wilderness, parks, or similar designations. The relative extent of these categories for all U.S. forestland is illustrated in figure 2-1. Virtually all reserved forests are in public ownership. Timberland accounts for an estimated 504 million acres (67%) of private forestland and noncommercial forestland for 191 million acres (26%).

Table 2-1.
Comparison of land area and forest area by U.S. region (000s of acres)

Region	Total Land Area	Total Forest Area	Forest as % of Land
Intermountain	547,918	135,499	24.7%
Alaska	365,039	129,131	35.4%
Pacific NW	104,054	48,481	46.6%
Pacific SW	103,934	39,011	37.5%
Great Plains	194,299	4,232	2.2%
North Central	286,674	83,108	29%
Northeast	128,816	85,380	66.3%
South Central	387,104	123,760	32%
Southeast	147,419	88,078	59.7%
Totals	**2,265,320**	**736,680**	**32.5%**

Source: Powell et al. 1993

Generally speaking, public forestlands are less productive for timber than private forests, as discussed more fully later in this section. Analyses of forestland and timberland areas for each state and region, organized by ownership type (IPF-NIPF), are presented in appendix B table B-1.

The data presented in this section are drawn primarily from the USDA Forest Service's RPA estimates for 1997, and secondarily from Powell et al. (1994). There are many more forest types than those categorized in the national forest resource assessment, which by tradition has been organized around major commercial timber species. These fail to reflect the impressive diversity of American forests, which range from the black spruce–dominated boreal forests of Alaska, to the distinctly different northern and southern Rocky Mountain forests, to the many unique oaks across California's threatened rangelands, to the incredible diversity of the mixed mesophytic Appalachian forests, and the increasingly rare wetland forests of the southeastern coastal plain. In the interest of simplicity, for this brief overview of U.S. forests we will utilize the USDA Forest Service types. However, there is considerable literature available for those interested in the breadth of forest cover in the United States. One reference with a wealth of ecological detail is *North American Terrestrial Vegetation,* edited by Barbour and Billings (1988). Further information on species diversity and some elements of stand structure can be obtained from the Forest Inventory and Analysis (FIA) units of the USDA Forest Service

Figure 2-1.
Classification of forestland in the United States

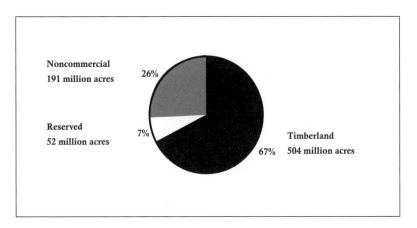

at regional research stations. The RPA data can be accessed on the World Wide Web through http://fia.fs.fed.us.

A Snapshot of Eastern Forests

The USDA Forest Service tracks ten forest types in this region (comprising the Northeast, North Central, Great Plains, Southeast, and South Central areas). Covering a total of about 384 million acres, their respective areas are presented in figure 2-2. Hardwoods predominate, accounting for about 70% of eastern forests. Eastern forests are 85% privately owned. Nonindustrial ownerships are 80% hardwood in composition, while industrial owners hold proportionately more softwood lands (47% of industrial acreage). Public forests in the East are more limited than in the West; half are federal (mostly national forestland in the more mountainous areas of New England, the Great Lakes states, and the South Central region) and half state forests (notably in Pennsylvania, Michigan, Minnesota, and Wisconsin). As in the West, the more productive soils of the coastal regions are almost entirely privately owned.

The 130-million-acre oak-hickory forest, found throughout the southern regions as well as the southern half of the northern regions, is the most extensive. Close to 90% of this forest type is private, mostly in nonindustrial ownership. The maple-beech-birch forest is the second most common forest type, with 54 million acres in the Northeast and North Central regions, almost entirely in both industrial (13.7%) and nonindustrial (71%) private ownership. The extent of this latter forest has been expanding through silvicultural practices that favor it. Since the 1992 RPA, this forest type covered an additional 8 million acres.

The oak-gum-cypress forest type is found on the bottomlands in the Southeast. It covers about 30 million acres. After generations of conversion to agricultural use, the area occupied by this type has been relatively stable in recent years. Another bottomwood forest type found in both the North and the South is the elm-ash cottonwood forest, whose 13 million acres often include wetlands. It too is greatly diminished in extent through loss to agriculture and other uses.

Another major southern hardwood type is the oak-pine forest, which, as Powell et al. (1994) note, resulted from the selective harvest of natural pine stands. This forest type has been declining, primarily through conversion to loblolly pine plantations. Its extent is estimated to be about 34 million acres.

In the North Central region, the aspen-birch forest is a significant type, with almost 18 million acres (80% of all aspen-birch forests). These are pioneer forests that colonize lands disturbed by fire or clear-cuts, or that take over abandoned agricultural lands. While of significant wildlife value,

Figure 2-2.
Forest types on unreserved forestland in the East, 1997

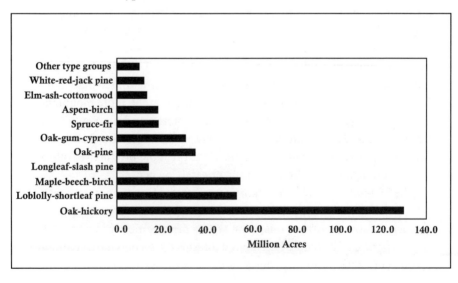

especially for white-tailed deer, aspen has become a major commercial fiber species.

In general, softwoods cover a much smaller area than hardwoods in the East. The loblolly pine has the second greatest extent of the major forest types, with 52.5 million acres. This represents an increase of almost 7% from 1992 to 1997, according to RPA data. The loblolly and other short-leaf pines occur almost entirely in the South and are preferred plantation species. This forest type represents more than half the conifer forests of the East. Longleaf-slash pine forests now cover only 13 million acres or about 2% of their original area. The remaining longleaf pine forest is concentrated in Georgia and northern Florida, with additional acreage primarily in North Carolina.

Northern softwoods are mostly of the spruce-fir type (17.4 million acres), followed by white-red-jack pines (11.6 million acres). Within these forest types, there can be considerable regional variation of species. For instance, in the white-red-jack pine type, white pine is the dominant species in the Northeast while red and jack pines dominate in the North Central region.

A Snapshot of Western Forests

Western forests (including those in the Intermountain and Pacific Coast regions) primarily consist of softwoods. Only about 12% of western forest acreage is hardwood. Figure 2-3 presents the coverage of the major western species. Most of these forests are publicly owned (67% in aggregate). Public ownership ranges from a high of 85% in the Intermountain region, with its less productive timber soils, to a low of 44% in Washington, where state trust lands are a significant ownership type. Of the private lands, industrial owners own proportionately more softwood lands (86% of industrial acreage) than do nonindustrial owners (44% of NIPF acreage).

Of the 362.5 million acres of western forest, almost 124 million are in black spruce stands in Alaska (along with several other minor conifer species) and in pinyon-juniper woodlands in the Intermountain region. Most of these are on public lands. One-third of pinyon-juniper forests and a scant 5% of black spruce and other conifers are in nonindustrial private ownership. These types have limited commercial value but make major contributions to wildlife habitat, biodiversity, and watersheds.

Western hardwoods cover about 42.5 million acres. There are distinctly varying hardwood ecosystems across the West: Aspen is the most abundant hardwood in the Intermountain region with oaks of various types predominating in California and red alder in Oregon and Washington. Although they are a relatively minor component within western forests, hardwoods have great ecological value. Historically these western hardwoods were not considered to have much commercial value beyond use for fuelwood. However, as in the northern regions, aspen has become a major pulp species. Red alder, long considered an annoying weed in the Douglas-fir region, has recently experienced an explosion in its commercial value as a sawtimber species utilized in the furniture industry. Certain other hardwoods, such as black oak, madrone, big leaf maple, and tanoak, are also emerging as commercial lumber stocks.

The primary softwood forest types of the West are fir-spruce (69.7 million acres), Douglas-fir (41.8 million acres), and ponderosa pine (33 million acres). The fir-spruce forests are found across the West in middle to upper elevations. Douglas-fir is also wide ranging in the West except in Alaska, and dominates in the Pacific coastal region. Ponderosa pine is found primarily in the Cascades and eastern portions of Washington, Oregon, and California, as well as in the Intermountain area. Western softwoods are renowned for their strength and utility in the wood products industry.

Figure 2-3.
Forest types on unreserved forestland in the West, 1997

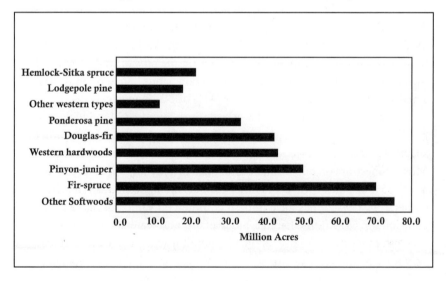

Lodgepole pine is another key forest type, with about 17.5 million acres, primarily in the higher-elevation Intermountain region. Hemlock-Sitka spruce forests comprise about 21.4 million acres, and are found primarily in the coastal areas of Oregon, Washington, and Alaska. Larch and western white pine are less extensive forest types found on the east side of the Cascades and in the Intermountain region.

Redwood is a small but unique forest type occurring in a narrow band of the coastal temperate rain forest from Monterey, California, to just north of the Oregon border. Ninety-four percent of the 916,000 acres of redwoods are in private ownership. Redwood and Douglas-fir can accumulate more biomass per acre than perhaps any other forest types globally, making them especially significant carbon sinks. They are also the most productive commercial species in the United States in volumes of timber per acre.

Forest Productivity

Site indices, which estimate timber volume growth, are commonly used to classify land according to its potential timber productivity. As they incorporate such factors as soil quality, topography, and climate, site indices can be useful in understanding ecosystem productivity as well. Site index is measured in cubic feet of growth per year or, alternatively, is represented by the height of the tree at fifty or one hundred years of age. Each U.S. forest region has a range of site index classes, as shown in figure 2-4. For instance, over half of the unreserved forest land in the South Central region is in the two highest site classes for timber production, a much different proportion than is indicated for the other regions shown.

According to Powell et al. (1994), the largest acreages of the highest site class are found in the eastern oak-hickory and loblolly-pine types, and in the western coastal Douglas-fir type. The Pacific Northwest and California contain the greatest combined acreages of the two highest site classes, mostly within the coastal temperate rain forests bioregion. (Seventy-nine percent of California's redwood subregion is rated at 120 cubic feet, the highest class.) The lower-site forests are either in the drier, higher-elevation western forests or at northern latitudes. These include Alaska's fir-spruce boreal forests and the pinyon-juniper lands of the Southwest. Owing to their relatively nutrient-poor soils and harsher climates, lower-site forests can experience greater difficulties in regeneration from disturbances.

In general, the most productive sites (more than 50 cubic feet/year) are privately owned, as indicated by the lower proportion of noncommercial

Figure 2-4.

Proportion of unreserved forestland in site classes for four regions

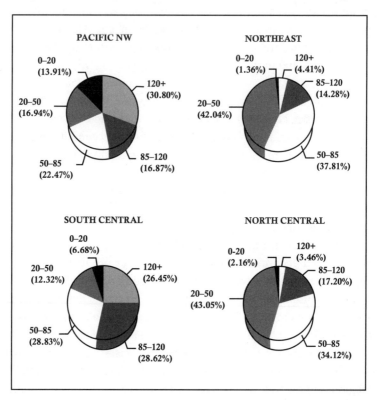

forests in private hands (figure 2-5). Of the 65.5 million acres of the highest-producing forest sites (site index above 120), more than 77% are private, with 15.6 million acres owned by industry and 34.6 million acres held by nonindustrial owners (the remainder of the lands in the all owners segment are in public ownership). Not surprisingly, industrial owners possess a greater proportion of higher site classes than nonindustrial owners, as discussed below.

Loss of Private Forest Extent and Function: Conversion and Fragmentation

Massive conversion of forests accompanied the settlement and development of the United States as land was cleared for agriculture and other

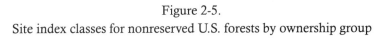

Figure 2-5.
Site index classes for nonreserved U.S. forests by ownership group

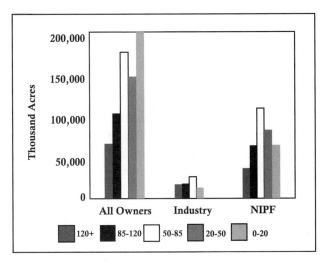

uses. Today, swelling population growth, increasing urban sprawl, and the creation of more and smaller forest parcels still erode our private forests, reducing both their extent and their ecological function. The area of private land in forest at any time is shaped by competing social and economic demands. Conversion arises when the economic value of alternative uses outweighs the economic value (among other values) of keeping the land as forest. This can occur when urbanization spreads into previously rural forest areas, increasing real estate development value. The relative value of alternative uses can also rise when the productivity, function, and/or timber value of a forest tract has been degraded by poor management practices or through parcel fragmentation. Together, forest degradation, fragmentation, and encroaching development can create a negative synergy that feeds further forest loss.

Forestland Conversion
It is uncertain how much remains of the forest that existed in the United States before European settlement. The USDA Forest Service estimates that today's forestland area is about 70% of the 1.04 billion acres of forest that existed in 1600. Other estimates put forest loss at 50% by World

War I, with some recovery of eastern forests since that time (NRC 1998). An estimated 307 million acres have been converted to other uses, mainly agriculture. More than 75% of this conversion occurred in the nineteenth century. Powell et al. (1994) note that between 1850 and 1920 more forest—190 million acres—was cleared by farmers than during the preceding 250 years. Clearing during this period averaged 13.6 square miles of forest every day for sixty years. Net forest loss, considering both conversion and regeneration of abandoned agricultural lands, diminished on a national scale during the twentieth century. Notwithstanding this relative stabilization of forest extent, lands that have recovered from previous historic clearing, such as the forests of New England, are distinctly different in composition and quality than they were at European settlement.

Clearing continues in metropolitan areas and other rapidly growing regions. For example, since World War II an estimated 24% of Oregon's commercial timberland was converted to agriculture or developed uses (Oregon Department of Forestry 1997).[1] The U.S. Department of Agriculture estimates that between 1945 and 1992, forestland extent declined by about 7%. While that may sound like a relatively small loss, it amounts to close to 40 million acres, an area greater than the size of Michigan (Daugherty 1995). During this same period urban/developed uses increased nationally by more than 285%.

Tracking changes in forest extent is made difficult by differing definitions used within the USDA (the USDA Economic Research Service, the USDA NRCS, and the USDA Forest Service) as well as USDA Forest Service classifications that change from decade to decade. One of the major sources for data is the NRI, produced every five years by the NRCS. The NRI tracks a number of factors including conversion among nonfederal land uses. While nonfederal land includes state-owned lands, the vast majority of it is privately owned. Changes in forestland area predominantly affect private lands.

---

1. The rate of forest conversion in Oregon slowed from 4% per decade in the 1960s to 2% in the mid-1970s to mid-1980s to 1.3% in the mid-1980s to mid-1990s (Zheng and Alig 1999). The decrease is coincident with Oregon's passage of the Land Conservation and Development Act in 1973, which established statewide land use regulation. This reduction has been achieved despite a booming state economy and a population growth rate 50% higher than the U.S. rate from 1960 to 1990. Whether this can be attributed to the LCDA is unclear (Kline and Alig 1999).

The 1997 NRI, according to revised data published in December 2000, shows an apparently substantial gain of 3.6 million acres in nonfederal forest extent during the fifteen years from 1982 to 1997. This increase is due primarily to conversion of pasture or cropland to forest by active replanting or by natural regeneration of abandoned land. However, in spite of the net gain of some 13.5 million acres of forest from these farm lands, the NRI estimates that during this same period about 10 million acres of nonfederal forestlands were converted to developed uses. That is an area of forest about twice the size of Massachusetts. The rate of conversion of forests to developed uses during this period is accelerating. During the last five years measured (1992 to 1997), 70% more forests were converted than in the previous ten years, running at about 950,000 acres per year of forest lost. While many rural acres may shift back and forth between forest and agricultural uses, loss to development is permanent.

It is important to recognize that even if total forest area is holding steady, or gaining slightly, the forest itself can be changing character significantly. Forests lost to other uses are often different in age, composition, quality, and ecological value than the forests gained. For instance, there are great differences between new acres of pine or poplar plantations on former agricultural land and lost acres of native hardwoods in natural stands. Further, the rate of urbanization is such that even the largest tree-planting effort in U.S. history, in which the Conservation Reserve Program reforested 2.6 million acres of cropland between 1987 and 1992, could not make up for the 3 million acres of forest converted during the same period (Alig et al. 1999).

The increasing rate and scope of urbanization is affecting not only forestland but agricultural land as well. This indirectly pressures forests further by reducing the supply of agricultural land that can be planted as replacement forest (Alig 2000). Twenty-three million acres in the South that are suitable for both timber and agricultural uses have shifted back and forth in use, depending on the markets. With global markets shifting agricultural supplies, it is very difficult to project what will become of these lands in the next decades. However, it is likely that other developed uses will increasingly provide competitive returns and that more agricultural land and forestland will be lost.

National statistics mask significant regional conversion of forests. While certain states, notably Texas, Missouri, Mississippi, and New York, are seeing large increases in forestland due to the replanting or recapture of agricultural lands, some major forested states, in particular North

Table 2-2.

States with net forest loss of 2% or greater (1982–1997)

State	Acres Converted	% Total Forest
Nevada	61,000	16.7%
Massachusetts	326,300	10.6%
New Jersey	201,800	10.6%
Colorado	315,300	8.4%
North Carolina	1,001,000	5.9%
Delaware	18,900	5.1%
New Hampshire	196,100	4.8%
Arizona	187,400	4.3%
California	564,600	3.9%
Maryland	85,000	3.5%
Connecticut	63,900	3.5%
Rhode Island	14,000	3.5%
Florida	326,800	2.5%
Washington	262,800	2.0%

Source: USD NRCS 1999

Carolina and California, are losing hundreds of thousands of acres to other, primarily built, uses. Table 2-2 shows the fourteen states that experienced serious net losses of forestland between 1982 and 1997, according to the revised 1997 NRI.

In addition, several other states experienced considerable local conversion of forests to nonforest uses during the same period, including Georgia (355,500 acres lost), Louisiana (194,000 acres lost), and Virginia (140,000 acres lost).

With the upsurge in real estate development and urban sprawl associated with the burgeoning economy of the 1990s, rates of conversion accelerated in the period 1992 to 1997. For instance, Georgia lost more than five times the area of forest from 1992 to 1997 than in the previous ten-year period. While Georgia still has the most timberland in the United States, it now ranks third in annual rate of urban development (Alig 2000). In California and Florida, the NRI shows that more forest was lost during the mid-nineties than in the previous decade. Forest loss appears to be growing exponentially in California, where more forestland was lost in this fifteen-year period (1982–1997) than during the thirty years from 1950 to 1980. Similarly, virtually all of the forests converted in Virginia, Nevada, and Massachusetts for the fifteen-year period (1982–1997) occurred in the last five years measured.

Many factors influence land use conversion, including demography, physiography, and culture (Alig 1986). A variety of studies suggest that forces fueling development include increasing overall population, increasing older population, increasing personal income, decreasing family size, increasing number of small households, and movement of populations to rural areas and warmer regions (e.g., Alig 1986; Parks and Murray 1994). Increasing income alone has been shown to be a key driver of urbanization and related forest loss, as more investments are made in building and other development (Alig and Healy 1987; Zheng and Alig 1999). Topography is another important factor: relatively flat or gently sloping forestland is most often subject to conversion. Major transportation corridors are vectors of conversion.

Owing to these forces, more than 1 million acres per year of forestland in the South are being converted to urban uses (NRCS 1999). The region's population grew by 84% between 1950 and 1990, compared with the national rate of 60% for this period. Populations doubled or tripled in the expanding urban areas of the Piedmont Crescent, Florida, and other coastal areas. Portions of the Pacific region, including California, Oregon, and Washington, experienced similar explosive rates of growth.

The NRC confirms that the United States is again in a period of accelerating private forest loss. They expect a decline of about 5% or another 20 million acres by 2020 (NRC 1998). On the other hand, the USDA Forest Service sees a more gradual loss. According to the USDA Forest Service's most recent study of changes to timberland area in the United States (Haynes 2001), timberland area will decrease by only 3% between 1997 and 2050. While the degree of projected loss is subject to debate, its likely source is not: most of the expected conversion will come from the change in nonindustrial forest ownerships to residential uses. Notably, however, industrial ownerships are expected to decline as well, for reasons discussed earlier, continuing a trend since 1987 (Alig 2000).

The USDA Forest Service expects the greatest losses in the next fifty years to come from the Pacific region. There, more than 12 million acres of forestland, mostly NIPF forests in California and western Washington, are projected to be converted. Close to 20% of NIPF timberland in California is expected to be lost to development. The South is anticipated to lose more than 8 million acres of NIPF timberland to urbanization, primarily in Georgia, Florida, and Alabama. These western and southern losses track the projected population growth in those regions of 15.7% and 12.9%, respectively, which for the 1990s was well above the national average of 9.6% (Alig 2000). As the forests of New England are within one day's drive of 70 million people, the Northeast will continue to lose timberland to second-home and recreational development. The Great Lakes states will continue to experience similar trends. In the Intermountain region, ongoing residential subdivision will eat away at lower-elevation private forests in expanding metropolitan areas. As with California, many smaller and mid-sized cities are expanding from their valley floors and development is moving up the forested slopes. Livable cities near national parks and other public lands are booming: from 1950 to 1990, the population of Estes Park, Colorado, grew by 104%; of Jackson Hole, Wyoming, by 260%; and of Missoula, Montana, by 91% (U.S. Dept. of the Interior 1998).

Forestland Fragmentation

Another element of forest loss is fragmentation. Fragmentation occurs at two basic levels: division or parcelization of larger, single-ownership forest tracts into smaller parcels with diverse ownerships, many of which become more developed; and reduction in forest patch size, combined with isolation among patches. Parcelization often yields smaller forest patches, as can forest management activities, such as clear-cutting. Fragmentation of forest and other habitats is "one of the greatest threats to biodiversity worldwide" (Burgess and Sharpe 1981; Noss 1983, 1987; Harris 1984; Wilcox and Murphy 1985; NRC 1998). We will discuss the impact of fragmentation on forest habitat values in more depth later in this section. When the effects of fragmentation are taken into account, the amount of functional forest area in any given region can be significantly diminished.

The increase in smaller forest parcels is driven by the demographic trends noted above, yielding increasing densities of people in forest areas. This trend is not revealed in aggregate numbers of forestland extent that are derived from the application of the USDA Forest Service definition of forestland (i.e., 10% stocking of trees of any size) to the U.S. landbase. This definition encompasses lands that may contain little forest, or may be covered with vegetation types such as chaparral. Therefore, it is necessary to look at other data to observe the fraying of the forest fabric. Nationally more than 32 million acres (an estimated 8% of total private forest area) are in parcels smaller than 20 acres. As discussed in chapter 1, smaller forest parcels are increasing at an accelerating rate. On average, almost 2 million acres of forestland per year—an area about the size of Yellowstone National Park—were broken up into parcels smaller than 100 acres between 1978 and 1994 (Birch 1996).

While still accounted for statistically as forest, these smaller parcels are often essentially residential and have been converted de facto from functional forest ecosystems. As forest areas urbanize, the capacity of the forest to maintain its functions, whether as wildlife habitat or for timber availability, becomes diminished. Therefore, "[t]he extent of forest may serve as a misleading indicator for the relative scarcity of the services provided by forests . . . and no-net-loss in timberland may mask substantial declines in services rendered . . . " (Wear et al. 1998).

In addition to parcelization of the more accessible rural forest areas, in the last twenty-five years urban areas that were formerly surrounded by forests have grown to encompass and diminish those forests. In an analysis of metropolitan statistical areas in the Southeast by the University of Georgia, in 1991, researchers found that these metropolitan counties contained 26% of the region's timberland acreage (28 million acres) and a comparable amount of the timber inventory (DeForest et al. 1991). Researchers in the Urban Ecosystem Analysis Program of American Forests compared satellite imagery of Atlanta, Georgia, for 1972 and 1993, and that of Washington State's Puget Sound for 1972 and 1996. For Atlanta, they found that almost two-thirds of the local forest had been absorbed by the metropolis. In Puget Sound, urban growth had doubled and a third of the forest cover had been lost (Dietrich 1999). Washington State has prepared for more growth by allowing counties to rezone 25% of forestland for future development. Virginia's Fairfax County lost 40% of its forest during the same period. The state of Virginia, in fact, has determined that 20% of its forest—3.1 million acres—is now urbanized (Liu 1997). Similar patterns are likely for every forested metropolitan area.

David Wear of the USFS Southern Research Station developed a population density model to predict the availability of timber and, as a corollary, the degree of forest fragmentation and impairment of forest functionality. In a study of the five counties that surround Charlottesville, Virginia, Wear and his colleagues found that as the density of population increased, the probability of the forest functioning as timberland decreased. At forty-five people per square mile (psm), the chance was 50%; at 70 psm the probability was 25%; and at 150 psm the likelihood was zero (Wear et al. 1999).

In California, the state Department of Forestry considers forestland functionally impaired through parcelization when there is one house per 40-acre parcel. On this basis, in addition to losing one-third of its original forest extent to other uses, another one-third of California's forest has been significantly fragmented, according to a Geographic Information Systems (GIS) analysis conducted by PFT in 1997.

Cover Type Conversion

Another form of conversion is that of cover type, where one kind of forest changes to another through management or other human influences. For instance, forests dominated by softwoods can be converted to hardwoods, or vice versa. Natural forests can be converted to plantations. For instance, in the draft 2000 RPA Assessment of Area Change, Alig noted, "Over the past 50 years, the largest changes in the private forests of the United States have been the substantial decrease in the area of natural pine and the rapid increase in the area of planted pine in the South." When cover type conversion due to poor forest management leads to reduced forest health or short-term economic productivity, this can yield increased fragmentation and subsequent conversion to other nonforest uses.

Although all forest ecosystems have experienced some degree of loss and degradation and many are under threat of substantially more, researchers have highlighted certain forest ecosystems in the United States as in being in peril of complete loss. Reed Noss and Robert L. Peters (1995) worked with the Defenders of Wildlife to identify what they felt were the most endangered ecosystems based on four factors: dramatic diminishment in area since European settlement; small and fragmented current area; relatively high numbers of imperiled species; and continuing threat to these species' existence. Twelve forest ecosystems were included:

- Southern Appalachian spruce-fir
- Longleaf pine forest and savanna
- Eastern grasslands, savannas, and barrens (including midwestern oak savannas)
- Northwestern grasslands and savannas (including Oregon's Willamette Valley oak savannas)

- Southwestern riparian forests
- Hawaiian dry forests
- California riparian forests and wetlands
- Old-growth eastern deciduous forests
- Old-growth Pacific Northwest forests (California, Oregon, Washington)
- Old-growth red and white pine forests (Great Lakes states)
- Old-growth ponderosa pine forests
- Southern forested wetlands

The World Wildlife Fund (WWF) has similarly identified globally out-standing ecoregions believed to require immediate protection of remaining habitat and extensive restoration as part of a comprehensive strategy for conserving global biodiversity (Ricketts et al. 1999). Included are twelve U.S. forest ecosystems the WWF considers either critical, endangered, or vulnerable owing to a variety of factors:

- Hawaiian moist forests
- Hawaiian dry forests
- Appalachian/Blue Ridge forests
- Appalachian mixed mesophytic forests
- Southeastern conifer forests
- Southeastern mixed forests
- Pacific temperate rain forests: Central Pacific and Northern California
- Klamath-Siskyou conifer forest
- Sierra Nevada conifer forests
- California interior chaparral and woodlands
- California montane chaparral and woodlands

Timber Resources, Productivity, and Harvest

Forests are highly valued economically for their capacity as timberland, which yields a variety of wood products. Timberland is forestland that is biologically capable of producing timber yields of 20 cubic feet per acre per year or more, and which is not reserved from timber harvest by legislation or regulation.

Therefore, in the following statistics on timber resources, the land base in question is the 504 million acres of timberland, not the full 747 million

acres of forest. Almost three-quarters of timberland is in private ownership (13% in industrial ownership and 58% in nonindustrial). The
amount of timberland in the United States decreased by 4% (19.3 million acres) from 1952 to 1991, largely because of public acquisitions or
set-asides for wilderness or parks. However, some of the loss is attributable to conversion to other uses. The USDA Forest Service has projected a comparable loss in timberland between 1992 and 2040 (Alig
and Wear 1992). Given the factors discussed above that are driving
increasing parcelization and urbanization of prime timberland in many
regions, this projection may potentially underestimate the acreage that
will no longer be available as high-functioning forest for timber or other
benefits.

Most private forests (about 85%) fit the timberland category. In
mountainous regions such as the West or the eastern Appalachian
chain, private timberlands are commonly found on the lower, gentler
portions of the landscape, with public forests occupying the more
mountainous and remote areas. As a general rule, inherent soil productivity is inversely related to the steepness of slope and to altitude,
so private forests are usually some of the most productive timber-
growing soils in the region.

Timber Stocks and Growth
In the FIA of 1997, U.S. forests were estimated to hold 906 billion cubic
feet of timber, of which 92% was considered to be merchantable growing stock, that is, sound trees with good form for wood products. About
60% of the growing stock is on private land. For the nation as a whole, the
amount of standing timber volume increased 33% from 1952 to 1991,
mostly in hardwoods. This increase is the result of recovering forest inventories in the East after depletions earlier in the century. Inventories actually fell in the Pacific region, primarily because of the continuing removal
of old-growth forests as well as creation of new parkland and wilderness
areas. Figures 2-6 and 2-7 illustrate hardwood and softwood growing
stocks by ownership type and region.

Softwood Stocks
Softwoods account for almost 58% of U.S. timber growing stock, of which
68% is in the West. Douglas-fir accounts for 23% of all softwood volume,
followed by loblolly and shortleaf pines with 14%. Softwoods are the pri-

Figure 2-6.
Softwood and hardwood growing stock volume by ownership (1997)

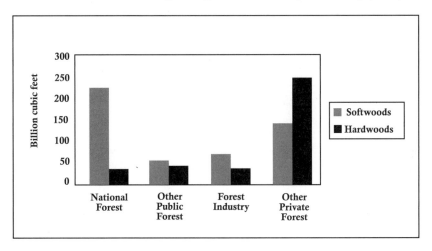

Figure 2-7.
Softwood and hardwood growing stock volume by region (1997)

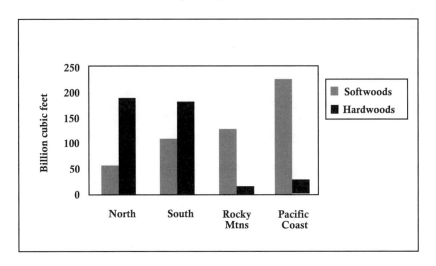

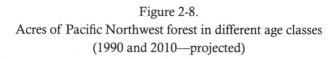

Figure 2-8.
Acres of Pacific Northwest forest in different age classes
(1990 and 2010—projected)

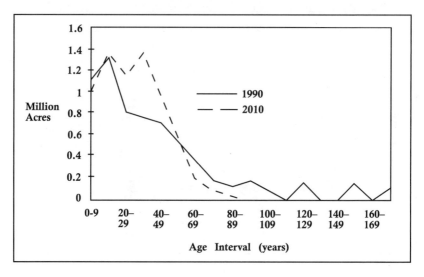

mary source of dimensional lumber and building materials, such as plywood, oriented strandboard, and other engineered wood products. High-quality logs are also major sources of veneer. Softwoods are the primary source of pulp for paper.

Softwood inventory overall grew by about 11% between 1952 and 1997. However, industrially owned standing timber inventories declined 14.4% nationally during the same period (falling almost 9% in the ten years from 1987 to 1997). In fact from 1952 to 1997, softwood inventory on indus-trial land in the Pacific Northwest fell by 42%—a particularly telling sta-tistic given that Pacific Coast forests account for 44% of all softwood volume (in spite of the relatively small area of timberland in the region). During this same period, NIPF inventories were the only ones to increase, growing by 52%. The USDA Forest Service predicts total private forest softwood inventories to increase by 40% over 1997 levels by 2050 (Haynes et al. 1995).

From 1953 to 1992 the volume of softwoods in larger sawtimber classes (with diameter at breast height, or DBH, greater than 19 inches) fell by 30% (Powell et al. 1994) With increasing intensity in softwood management and corresponding increases in timber harvests, average forest age and tree diameter size will continue to decrease. Although age is only one element of habitat suitability, this trend raises serious questions about ecological impacts to fish and wildlife habitat, forest carbon stores, and other forest values.

Pacific Coast forests exemplify the changing nature of softwood forest structure and age class distribution (figure 2-8). Although the primary commercial species in this region only begin to reach biological maturity when they are one hundred years old or more, the harvest age has decreased to forty to fifty years for greater financial efficiency. The average diameter of harvested trees shrank by 41% between 1976 and 1997, falling from a 27.5 DBH to one of about 16 inches. (This was *after* much of the old growth on private lands had already been harvested.) That average was projected to decline further to 14 inches by 2050.

Hardwood Stocks

Hardwoods account for 42% of growing stock. Oak species account for one-third of these volumes. Soft and hard maples are the next most abundant, with 17% of volume. However, current forest management is encouraging the expansion of maples in volume and distribution. High-quality hardwoods such as select oaks, hard maple, yellow birch, black walnut, and black cherry accounted for 40% of the eastern hardwood stocks. These hardwoods are very valuable for lumber, furniture, cabinetry, veneers, and specialty products, such as architectural woodwork and sporting goods. Lower-quality hardwoods are also used for a range of wood products, from fuelwood to pallets and, increasingly, for chips used in pulp and engineered building products, where new technologies allow them to replace relatively scarce conifers.

In the ten years between 1986 and 1997, hardwood inventories increased by almost 13%. In fact since 1952, hardwood volume has increased 91%, mostly through regrowth of previously harvested forests. Remember, nonindustrial owners hold 71% of hardwoods, and, as with softwoods, that is where the inventory increases are continuing. From 1986 to 1997, hardwood stocks fell 8% on forest industry ownerships in the South, reflecting continuing conversion to pine plantations. The USDA Forest Service projects hardwood inventories will rise only 23% from 1990 to 2050 as all private owners increase harvests and continue the move toward pine plantations. In addition, smaller nonindustrial owners are expected to convert more hardwood forests to residential uses (Haynes et al. 1995). In the South, 10% of hardwood forests are expected to be lost by 2040 (Alig et al. 1990).

In general, current management practices are tending to favor lower-quality hardwoods. One indication of this is that while overall hardwood stocks have grown, the average diameter of trees is shrinking. As of 1992 only 10% of hardwood volume was in trees with DBH greater than 20 inches, while the great majority of volume occurred in the 8- to 14-inch DBH size classes. In the Northeast, according to Powell et al. (1994), large-diameter hardwoods have been increasing as the resource recovers from earlier depletion.

Hardwoods are generally slower growing and have a tendency to dominate sites, and many are shade intolerant. These characteristics, combined with the objectives of industrial landowners, have caused a management-driven species shift toward softwoods on industrial forest lands. On NIPF lands, where timber-harvesting pressures are generally less, hardwood species have continued to grow.

Timber Growth

Owing to a combination of younger forests—which grow at a faster rate than older ones—and more intensive management on higher-site industrial lands, timber yields or net annual increase per acre (growth less harvest and mortality) on private lands rose through the 1970s and then leveled off. Timber growth rates for forest ownership types over the forty years from 1952 to 1992 are shown in figure 2-9. Note that the overall average annual net growth rate declined in the period because public forests were managed less intensively for timber—and retained more mature and old-growth stands—compared with private forests. Public lands are also generally lower site-class forests and therefore slower-growing by nature.

Today lands managed by the forest products industry annually grow around 20% more timber, on average, than those managed by nonindustrial owners. This is due to a combination of better average timber-producing sites brought about by selective acquisition of high-yielding lands, and the use of more intensive forest management practices, including prompt replanting after harvest; use of faster-growing species and cultivars; control of competing vegetation; and fertilization. Increasing rates of timber growth is essential to accomplishing a key

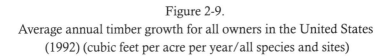

Figure 2-9.
Average annual timber growth for all owners in the United States
(1992) (cubic feet per acre per year/all species and sites)

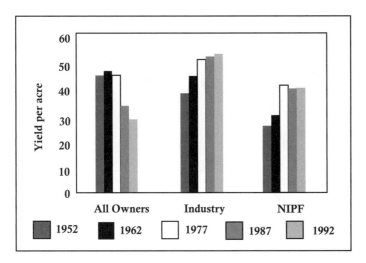

goal of intensive forest management: by shortening the cycle of growth to harvest, capital investments in growing stock are recovered as quickly as possible.

Increasing intensity of management and shorter rotations bring their own risks, however, potentially reducing long-term yields because of the increased removal of biomass and frequent soil disturbance (NRC 1998). Thus, the amount of fossil-fuel inputs required of intensive management may grow while producing diminishing returns.

Timber Harvest and Supplies

Timber harvesting is the major economic use of forests and has the greatest impact of any use on the structure of America's forest ecosystems. This section briefly reviews recent history, current status, and projections in timber harvest in relation to stocks, growth, and probable supplies. According to draft USDA Forest Service data, in 1997 total U.S. timber harvest exceeded 17.6 billion cubic feet, of which almost 90% was from private forestland, with IPF owners producing 29% and NIPFs producing 60%. National forests contributed 6.6%. Softwoods comprised 59%, and hardwoods comprised 41% of timber harvested

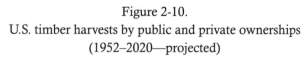

Figure 2-10.
U.S. timber harvests by public and private ownerships
(1952–2020—projected)

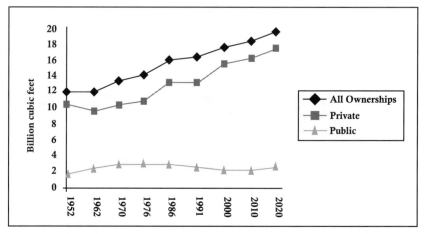

(Haynes 2001). The relative contribution over time of the major ownership types to U.S. timber harvests is illustrated in figure 2-10.

Over the last decade supplies have been shifting among sources because of a combination of factors, including reduced merchantable inventories of softwoods on industrial lands, cutbacks on commercial timber harvest from national forests, and increased harvest by nonindustrial owners in many areas. While total harvest declined slightly from 1991 to 1997, private harvest increased by 7%, or 1.02 billion cubic feet. Most of this increase came from nonindustrial owners, who increased their share of the cut by 10%, or 970 million cubic feet. Eighty percent of this increase came in softwood harvests, as NIPF owners responded to shortages on other ownerships and attendant higher stumpage prices. Harvests from national forests declined during this period by 66%, or 1.38 billion cubic feet, as federal forest management policy shifted in favor of protecting threatened habitats and providing other nontimber public benefits, especially in the Pacific region.

Timber harvests can vary considerably from year to year, depending on many factors. To better understand the impact of harvests on forest structure and future resources, it can be useful to examine the relationship of inventory growth to harvest (also known as the growth/drain or growth/removals ratio, although "removals" also includes wood removals, such as logging residues or noncommercial thinnings that do not result in forests products). If harvest is greater than growth, the ratio is less than 1. In this case, stocking is being depleted and forests ages are becoming younger on average. If harvest is less than growth, the ratio is more than 1. The forest stocks are growing and the forest is generally getting older. Looking at the relationships of harvest and growth to total inventory is also instructive: if the harvest/inventory ratio is greater than the growth/inventory ratio, then standing timber stocks are being depleted and forests are generally becoming younger. Any examination of these ratios should take into consideration the gross level of data available as well as the fact that these data are but a snapshot in time of conditions. Therefore, the ratios may not reveal changed conditions for other years, nor do they provide information on local conditions or particular species. Nonetheless, they are useful in understanding larger-scale trends.

As of 1996, the USDA Forest Service reported that the United States as a whole had a growth-to-removal ratio of 1.47 for all species (1.33 for softwoods and 1.71 for hardwoods) (Smith and Sheffield 2000). At the regional level, this ratio ranges from a low of .91 for softwoods in the South to a high of 15.26 for hardwoods in the Rocky Mountains.

As softwoods have long been the primary source of timber harvested in the United States, we will spend some time analyzing them. Table 2-3 presents the historical comparisons of softwood harvest to inventories and growth from 1952 to 1997, as well as draft 2000 RPA Forest Service projections through 2030 (Haynes 2001). Table 2-3E shows that overall softwood harvest in the United States trended below growth in the period from 1952 to 1991, although with increasingly more growth being captured for wood products. The national ratio of growth to harvest for all ownerships began to increase again because of logging reductions in national forests, even as harvests on private lands grew. Table 2-3C shows harvest exceeded growth on private forestlands at the national level through the 1990s, although the USDA Forest Service predicts relative harvest will decline slightly to just below growth over the next thirty years. This decline would allow some recovery of stocks, enabling inventories to rebuild and young forests to grow back into ages that allow for a return to higher levels of commercial harvest by the 2020s. In general, private forests in the United States now appear to be producing timber at or beyond their full capacity after significant inventory and production growth between 1952 and the late 1970s.

Table 2-3.
Softwood timber inventories, growth, and harvest for U.S. forestland
owners (1952–2030 as projected in 2001)

A. Industrial Forestland Owners

	Softwoods (billions of cubic feet)			Ratios		
	Timber Inven.	Ann'l Growth	Ann'l Harvest	Growth/ Inven.	Harvest/ Inven.	Growth/ Harvest
1952	70.7	1.8	2.7	0.03	0.04	0.67
1962	69.6	2.2	2.1	0.03	0.03	1.05
1970	69.5	2.5	2.8	0.04	0.04	0.89
1976	69.4	2.8	3.3	0.04	0.05	0.85
1986	67.4	3.0	4.0	0.04	0.06	0.75
1991	66.1	3.0	4.3	0.05	0.07	0.70
1997	61.4	3.3	3.9	0.05	0.06	0.85
2010	70.6	4.4	4.2	0.06	0.06	1.05
2020	79.1	4.7	4.8	0.06	0.06	0.98
2030	80.9	5.2	5.6	0.06	0.07	0.93

B. Nonindustrial Forestland Owners

	Softwoods (billions of cubic feet)			Ratios		
	Timber Inven.	Ann'l Growth	Ann'l Harvest	Growth/ Inven.	Harvest/ Inven.	Growth/ Harvest
1952	101.5	3.5	3.5	0.03	0.03	1.00
1962	110.7	4.4	3.0	0.04	0.03	1.47
1970	121.3	5.3	3.3	0.04	0.03	1.61
1976	130.4	6.0	3.5	0.05	0.03	1.71
1986	142.0	5.3	4.3	0.04	0.03	1.23
1991	148.2	5.0	4.4	0.03	0.03	1.14
1997	144.2	5.7	5.2	0.04	0.03	1.10
2010	168.6	5.9	4.6	0.03	0.03	1.28
2020	182.9	5.9	5.1	0.03	0.03	1.16
2030	193.9	6.2	5.5	0.03	0.03	1.13

C. Total Private Owners

	Softwoods (billions of cubic feet)			Ratios		
	Timber Inven.	Ann'l Growth	Ann'l Harvest	Growth/ Inven.	Harvest/ Inven.	Growth/ Harvest
1952	172.2	5.3	6.3	0.03	0.04	0.84
1962	180.3	6.6	5.1	0.04	0.03	1.29
1970	190.8	7.8	6.1	0.04	0.03	1.28
1976	199.8	8.8	6.8	0.04	0.03	1.29
1986	209.4	8.3	8.3	0.04	0.04	1.00
1991	214.3	8.0	8.7	0.04	0.04	0.92
1997	205.6	9.0	9.1	0.04	0.04	0.99
2010	239.2	10.3	8.8	0.04	0.04	1.17
2020	262.0	10.6	9.9	0.04	0.04	1.07
2030	274.8	11.4	11.1	0.04	0.04	1.02

D. Public Ownerships

	Softwoods (billions of cubic feet)			Ratios		
	Timber Inven.	Ann'l Growth	Ann'l Harvest	Growth/ Inven.	Harvest/ Inven.	Growth/ Harvest
1952	259.6	2.4	1.4	0.01	0.01	1.71
1962	269.4	3.0	2.2	0.01	0.01	1.36
1970	269.4	3.5	2.6	0.01	0.01	1.35
1976	267.0	3.7	2.7	0.01	0.01	1.37
1986	243.6	4.2	3.0	0.02	0.01	1.40
1991	235.6	3.9	2.6	0.02	0.01	1.50
1997	273.4	4.6	1.2	0.02	0.004	3.83
2010	315.6	5.2	1.3	0.02	0.004	4.00
2020	356.1	5.1	1.3	0.01	0.004	3.92
2030	394.0	4.9	1.4	0.01	0.004	3.50

Table 2.3. *(continued)*

E. Total U.S. Ownerships

	Softwoods (billions of cubic feet)			Ratios		
	Timber Inven.	Ann'l Growth	Ann'l Harvest	Growth/ Inven.	Harvest/ Inven.	Growth/ Harvest
1952	431.8	7.7	7.5	0.02	0.02	1.03
1962	449.8	9.6	7.3	0.02	0.02	1.32
1970	460.3	11.3	8.7	0.02	0.02	1.30
1976	467.0	12.5	9.5	0.03	0.02	1.32
1986	453.0	12.5	11.3	0.03	0.02	1.11
1991	449.9	12.0	10.7	0.03	0.02	1.12
1997	479.1	13.6	10.4	0.03	0.02	1.31
2010	554.9	15.5	10	0.03	0.02	1.55
2020	618.2	15.8	11.3	0.03	0.02	1.40
2030	670.9	16.0	12.4	0.02	0.2	1.29

Source: Haynes 2001

Examining private forestlands in more detail, table 2-3A shows that industrial forestland owners have consistently cut softwoods at rates well above the growth of their forests, even as growth rates have been increasing. Harvests as a percentage of inventory have also increased. Only sometime during in the first decade of the twenty-first century will industrial harvest briefly lag behind growth, due essentially to a lack of merchantable timber. Then the pattern returns, with harvest outstripping growth through 2050. This situation is especially striking, considering that forest growth rates are projected to reach a level close to three times that of 1952 during this period.

Following the boom days of the 1950s, nonindustrial forest owners consistently harvested softwoods below the rate of their timber growth. However, table 2-3B shows the rate of harvest has steadily increased since the 1970s, peaking in 1997 with harvest just below growth. This means that at the moment the average age of nonindustrial forests is broadly stable. It is also younger than in the past—which can be deduced by the rate of growth—but not as young as industrial forests. Looking forward through 2030, NIPF softwood inventories are projected to build once more as harvest ranges between 78% and 88% of growth.

Regionally, RPA data from 1997 show that private forests in the North had a softwood growth/harvest ratio of 1.08 overall; however, industrial owners harvested almost three times growth. Harvest on both classes of private ownerships in the South increased to what some would consider unsustainable levels, yielding a growth/harvest ratio of .96. The harvest/growth ratio for the softwood-dominated Pacific region was .91 for 1997. Industrial owners in this region on average cut about 30% above growth, although this represented a reduction in cut from 1991, when industrial harvest was nearly twice the rate of growth. Nonindustrial ownerships in the Pacific region had a more conservative ratio of 1.18. In general, however, the NIPF ownerships increased their harvests between 1991 and 1997, particularly those in the Douglas-fir or Pacific Northwest–West subregion, where the growth/harvest ratio declined from 1.03 to .86 during this period. In addition to increased harvests, the decline is the harvest/growth ratio for the Pacific region is also due to the reduced growth on private forests of Oregon and Washington, reflecting significant depletion of older stocks.

Softwood inventories in the South declined between 1986 and 1991 and then began a slow recovery. However, when looking subregionally within the South, there are signs of major erosions of softwood inventories in the upper Piedmont of Georgia and Alabama, reflecting the impact of urban development (Wear 1996). When Professor Frederick Cubbage of North Carolina State University examined the then-current FIA data, he found that removal exceeded growth in all major softwood-producing states and survey units. In fact, the eight core Southern states removed 723 million cubic feet more wood than had grown for the period. That is enough fiber to supply ten large pulp mills for a year. Inventories for states such as North Carolina, Virginia, and Tennessee that have growth exceeding removal cannot compensate for this excess harvest elsewhere (Cubbage et al. 1995).

Given what the USDA Forest Service characterized in 1990 as the "unprecedented" (Haynes 1990) softwood supply constraint that is expected to last until at least 2015 or 2020 (when the young plantations of the Pacific Northwest and South will come of harvest age), harvest of the relatively plentiful hardwood stocks has been growing strongly during the 1990s. Thanks to changes in technology, hardwoods can now substitute for softwoods in a range of products, especially those made from chips, such as pulp and oriented strand board. The proportion of timber harvested from hardwoods in 1991 was about 33% and increased to 41% by 1998. The USDA Forest Service projects hardwoods to increase their share of harvest to a high of 45% in the 2010s and then gradually fall back to about 40% by 2050. This increase is expected to come primarily from NIPF ownerships, as they increase their share of hardwood harvest from 72% to 83%, as industrial ownerships continue converting hardwood acres to softwood plantations.

During the late 1990s, hardwood harvest in the South and in the Great Lakes states exploded, with removals at more than twice the rate of growth reported for Minnesota and Wisconsin. The chip mills that have expanded rapidly in the more mountainous regions of the South are taking advantage of a relatively cheap, plentiful, and little-regulated hardwood resource. The accelerated chip-driven harvest impacts on Southern forests have alarmed many landowners and policymakers—even making the front pages of the *Wall Street Journal* and the *New York Times* in 1999 and 2000. Increased attention to the environmental impacts and potential resource depletion that such accelerated hardwood harvests are yielding could lead to new constraints on some logging practices.

Projected Trends in Timber Supplies

Any discussion of timber supply and harvest projections must be undertaken with great caution. As with any economic projection, many variables and assumptions are used in creating the resulting numbers. As the projection extends farther into the future, its reliability becomes more unlikely given the probability of unforeseen circumstances. This does not stop anyone from making projections and relying on them, to a greater or lesser degree, for planning purposes. The USDA Forest Service and private economists periodically issue projections that are the subject of great debate and often disagreement.

Keeping these caveats in mind, the USDA Forest Service projected in its draft 2000 RPA Timber Assessment that harvest would increase to 23.6 billion cubic feet by 2050 (with softwood harvest growing by 46% and hardwood harvest by about 24% from 1996 levels) (Haynes 2001). Overall growth in harvest is projected to be at a rate of 0.6% per year between 1998 and 2050, about half the rate of growth between 1965 and 1998. Per capita consumption in the projection period is expected to increase by 7.7%, while per capita harvest decreases by 6.5%. The gap between domestic supply and demand is expected to be bridged by increased imports of both wood and wood products, recycling, and further improvements in efficient processing of raw wood into products. Interestingly, the use of recycled fiber is expected to grow by 130%, twice the rate of increase of any other source.

Imports, primarily from Canada, are predicted to grow to as much as 40% of consumption in the decade from 2010 to 2020, as compared with 24% today. Imports decline again to more historic proportions as softwood supplies expand after 2015 with the growth of industrial plantations to harvestable age. After 2020, the United States becomes increasingly reliant on plantations for softwood harvests. Currently, plantations account for 16% of softwood harvested; by 2050 they are projected to supply 43%, due to a 70% increase in plantations acreage. By 2050, plantations in total are expected to cover 30% of softwood timberland area while providing 55% of softwood harvest. Forest industry plantation area will increase by 155% to 16.1 million acres, and nonindustrial plantation area will increase by 117% to 7.6 million acres. In addition, the USDA Forest Service predicts that after 2030, fiber derived from woody crops grown on agricultural land will grow to represent 6% of hardwood harvest.

By 2050, NIPF harvest is expected to account for 63.5% of the total, while industrial and public ownerships are anticipated to supply 27.5% and 9%, respectively. To put this in context, in 1991, NIPFs accounted for 54% of all timber harvested in the United States, and by 1997, NIPF share of harvest had grown to 60% in response to the rising stumpage prices and supply shortages in certain species and log grades as national forest and industrial timber supplies were curtailed. For instance, in Oregon, NIPF harvests grew by 328% since 1981 with 592 million board feet harvested in 1997 (Kiesling 1999).

These projections reflect a dramatic change in the historic relative weighting of timber supply sources, with nonindustrial share increasing by more than 25% in the period 1990 to 2050, over 1950 to 1990. Perhaps it is axiomatic that timber harvests follow the wood; that is, if and when there is merchantable inventory, it will be cut. With industrial stocks stretched thin, maintained at relatively young ages and correspondingly low levels of average stocking, in general the nation's more elastic source of supplies is found on the less intensively managed nonindustrial land.

Pulpwood (non-sawtimber smaller trees) used in paper and paperboard, as well as composite products like oriented strandboard, will increase from 53% to 57%, again illustrating the relative youth of growing stock on private forestlands. Sawtimber consumption, on the other hand, falls from 47% to 43%.

According to the USDA Forest Service, the proportion of total wood harvest from the South is expected to grow from the 1997 level of 56% to 58% in 2020 and 60% in 2050. At the same time, harvests in the Pacific Northwest will fall from 19% of the U.S. total in 1997 to 16% in 2020, and 14% in 2050, due primarily to relatively greater abundance of supply in the South (especially among NIPF owners) and reduced merchantable sawtimber on industrial forests in the Pacific region. For instance, forest industry stocking levels are expected to decline by another 36% in California by 2050.

In fact, 98% of the projected increase in softwood harvests is expected to come from private timberlands in the South. To secure these increases, more and more forest area will need to be harvested more frequently, given the low volumes of merchantable wood per acre. "There are indications that the projected increase in Southern harvest will be faster than inventories can sustain and without changes in underlying assumptions, result in reductions (in harvest or inventory or both) beyond 2050," notes the draft 2000 RPA Timber Assessment. They project that removals will reach almost 115% of growth in 2045.

Given the growing primacy of the South as the "wood basket" for commercial processing, there are profound concerns about the sustainability of Southern forest resources, which are overwhelmingly private in ownership. This concern fueled the creation of the first interagency effort to assess forest ecosystem sustainability. Led by the USDA Forest Service, the "Southern Forest Resource Assessment" joins the U.S. EPA, U.S. Fish and Wildlife Service, Natural Resource Conservation Service, Tennessee Valley Authority, and state forestry agencies together to better understand in a more integrated fashion the status of the region's forest ecosystems and capacity. Begun in 1999 with a two-year time-frame, they intend to use current data to look at the status of, and relationships among, forest-age class distributions, species composition and conversions, fragmentation and nonforest conversion, timber supply, silviculture, harvesting intensity reforestation, processing technology, biodiversity, threatened and endangered species, wetland impacts, water quality, and supplies. Such a comprehensive assessment would be welcome not only in the South, but in all forest regions.

This discussion of timber supply is not complete without considering factors other than sheer volumes of standing timber inventory or the interplay of complex market forces that dictate the actual availability of wood for harvest. As discussed further below, these limiting factors include environmental and physical constraints, demographic trends, fragmentation of parcels, public opinion, and landowner objectives (Cubbage 1995). The resulting available timber supply may be anywhere from 10 to 75% less than would otherwise be projected. In fact, for the first time, in the 2000 RPA Timber Assessment the USDA Forest Service discounted inventories to generate an estimate of "available inventory" from which they project future harvests. They estimate that currently 7% of private timberland in the South is unavailable for harvest on this basis (including almost 11% of NIPF ownerships in the Southeast subregion), and they project that this percentage will double over the next fifty years. The combination of such "reserved" or inoperable private timberland plus the amount of inventory too young to harvest reduces available inventory to a range of 50 to 58% of total inventory between 1995 and 2045.

Environmental constraints include limits on harvest to protect threatened and endangered species, wetlands, and water quality. Physical constraints that limit timber harvest operability include steepness of slope, soil stability, and other characteristics. Demographic trends include the increasingly urban nature of where timber inventories are located, a context that tends to limit harvest due to the higher amenity value of the trees for landowners and other influences. Any one of these impacts can sharply reduce the availability of timber for harvest.

A number of studies have confirmed the findings in the draft 2000 RPA Timber Assessment regarding the likely result of these constraints. One analysis discounted then-current FIA hardwood timber projections based on issues of operability, reducing potential volumes by 75% in mountainous areas of the South, by two-thirds for the Piedmont, and by 60% for the Southern Coastal Plain (Araman and Tansey 1990). Cubbage recounts a similar study by the North Carolina Forestry Association in 1993 that showed 38% of timberland in the state would be unavailable for harvest, accounting for about 25% of hardwood and 10% of softwood inventories. When metropolitan timberland was also eliminated, it suggested that only about half of standing timber would actually be harvestable.

The analysis of five central Virginia counties by Wear et al. described earlier in this report, found that the effect of increasing population density reduced the area of commercial timberland likely to be harvested, and therefore the growing supply, by roughly 40%. Their research found that the transition between rural and urban land use occurs where population density is between 20 and 70 people per square mile. Using their model, they estimate that for each 20% increment in population growth, timberland area drops by 4% (1999).

In addition to the various physical and regulatory constraints on the actual availability of timber, the very nature of NIPF ownerships makes it difficult to predict future levels of timber harvest. The character of NIPF ownerships also suggests that the recent increase in their relative contribution to supply may be difficult to sustain, much less continue to grow. First, as we have seen, NIPF forests are those most at risk for continued fragmentation and conversion to other uses. Therefore, the timberland base in question is shrinking. Second, NIPF owners are hugely varied in their objectives for their properties. The potential for widespread adoption of a more intensive harvest regime by those who do not already practice industrial-style forestry is probably limited. Medium and larger NIPFs are more likely to expand their harvesting than smaller landowners. However, even they tend to own their forests for multiple objectives, including ecological and amenity-oriented ones, and therefore many may be less inclined to maximize wood production regardless, as compared to industrial owners. Last, unlike industrial owners, NIPF owners in general have not shown themselves willing to invest in planting, thinning, and other stand improvement treatments that increase inventories and growth rates without major public subsidies to defray the up-front costs.

If the level of future harvest projected by the USDA Forest Service is indeed sustained, America's private forests will also be sustained in a considerably younger and simpler condition compared to their biological capacity. Regardless of the faith one places in any set of projections of supply and demand, as the U.S. population grows by 95 million or more people in the next fifty years, the future holds intensifying demands for timber production from private lands during a period of intensifying competition among uses and desires for that forestland. The ability of forests to successfully supply the commodity and noncommodity goods and services desired by society is challenged by growing populations, increasing urbanization of forest regions, a booming economy and attendant real estate construction sector, and increasing demands for protection of the ecological values of forests.

Forest Health Concerns

Forest health depends on the vitality of forest ecosystems. The USDA Forest Service's concern for the health of U.S. forests has resulted in creation of a Forest Health Monitoring Program, a multiagency cooperative effort to gather data, develop forest health indicators, and issue national and state-level reports. The program's data show an increase in timber mortality of 24.3% from 1986 to 1991. Determinations of declining forest health are derived from a complex of factors including loss of biodiversity, forest fragmentation, degraded water quality, fire suppression, presence of invasive exotic species, and air pollution. In this section we will focus on the latter three concerns.

As forests are interrelated systems, one aspect of forest impairment is likely to be rooted in other elements of the system. Invasive exotic vegetation thrives in disturbed and fragmented forests, moving along road systems. Insect outbreaks can be a natural systemic response to crowded stands that lack vigor, itself a result of successful fire suppression. For instance, outbreaks of defoliating insects such as the western spruce budworm have become more extensive and of longer duration because of the homogeneous, overstressed condition of young, overstocked forests such as those on the east side of the Cascades in the Pacific Northwest (U.S. Dept. of the Interior 1998). However, endemic diseases and insects are part and parcel of the forest ecosystem and essential to its functioning, assisting in decomposition, nutrient cycling, habitat creation, and successional processes (DellaSala et al. 1995).

Conversely, diverse forests with more complex structures and age classes are often less susceptible to widespread outbreaks of disease or insects. "Homogeneous plant communities, whether naturally occurring ecosystems, forest plantations or agricultural monocultures, are more susceptible to outbreaks of pests and diseases, including exotic organisms, than more heterogeneous communities" (Ewel et al. 1999).

Fire Suppression

Fire is both an essential aspect of a healthy forest ecosystem and a widely feared threat to private forest investment, as well as to forest communities. Government involvement in private forestry originated with fire suppression activities. With consistent effort over the twentieth century, the average area of forest burned annually by wildfires has been reduced from 20 to 50 million acres to 2 to 5 million acres (Powell et al. 1994). This very successful effort allowed millions of acres of cutover forest to regenerate and provided a higher degree of security for private investment in forests. Fire suppression has also allowed for more residential development in forest regions, thanks to public confidence and lower insurance rates. However, the unintended negative impacts to forest health of this successful public policy implementation now haunt forest landowners in every region.

We now have very different forests than those encountered by European settlers; fire suppression is one of the major contributing factors to that change, combined with the impacts of fragmentation and certain silvicultural practices. Many forests are not only relatively young but also densely stocked and less vigorous. This has become the most common forest health problem, especially on NIPF ownerships, owing to fire suppression (Oliver et al. 1997; Sampson and Decoster 1998). Forests in many parts of the country contain more shade-tolerant species because they were previously controlled through relatively frequent, low-intensity ground fires. For instance, the Southwest has experienced an 81% increase in mixed conifer species, with understory species such as Douglas-fir and white fir taking over from the previously dominant (and more fire-adapted) ponderosa pine. Fire suppression is one of the major threats to the existence of the last remaining longleaf pine forests in the Southeast because it permits encroachment of invasive hardwoods. Within a few years of fire suppression, longleaf pine forest structure is so altered as to render it unsuitable for threatened species, such as the red-cockaded woodpecker. Where agriculture has been abandoned, as in much of the East, the returning forest is composed of species whose seeds spread readily and that thrive in the absence of fire. Further, fuel loads from the forest floor to the canopy are building in many regions, especially in more temperate forests. (Fuel loads build more slowly in cooler, wetter, and/or higher-elevation forests.) Paradoxically, as we saw with the dramatic increase in forest fires during the summer of 2000, the increase in forest fuel loads brought about by fire suppression in turn increases the risk, if not the likelihood, of catastrophic fires.

Trees, like all living species, produce more young than their environment can support. The excess fall prey to some control mechanism, and the rest survive to become the reproducing population. In many temperate forests, particularly those that were adjacent to grasslands, agricultural lands, or other areas of human activity, fire was historically the primary ecological process that maintained tree populations within healthy numbers. Before European settlement, Native Americans used fire as their main land management tool, and the resulting frequent, low-intensity fires were the dominant factor shaping many forests such as pine or oak savannas. The oak-hickory forest type became dominant in the East because of fire. With suppression, hard maple is succeeding oak. Fire probably burned 5.5 to 13% of California's total area each year in pre-European times (U.S. Dept. of the Interior 1999). Fire cycles vary for different forest types, but in general, low-intensity fires regularly cleared out woody understory vegetation and kept shade-tolerant species in check. High-intensity stand replacement fires occurred infrequently—every 100 to 350 years.

Fire suppression has not only changed our forests, it is changing the kinds of fires we are having. Fuel loading and more homogeneous structures have increased the size and intensity of individual fires, increasing mortality of vegetation and animals. Hotter, longer-burning fires hurt the soil and make regeneration of vegetation more difficult. In the growing urban-rural interface in forest areas, fires are also causing greater residential damage than in the past, and substantial public resources are being spent on protecting homes built in previously remote forest areas.

As the United States was settled, fire suppression became the management objective. Consequently, most of America's forestland has been deprived of a basic ecological process for 100 to 350 years. Clearing for agriculture and harvesting timber became the most common ways in which trees were cycled through life and death. But there are critical differences. Fire tends to kill the smaller, weaker, or less site-adapted individuals in the vegetative population, leaving green trees and other organic matter. Fire is also typically patchy in its impact, burning more intensely in one area while avoiding an adjacent area altogether. Clearing takes out most, if not all, the tree and brush species, while timber harvest typically takes out the largest and healthiest trees, and the resulting regrowth can become dense thickets.

Forests without normal fire cycles require some other form of population control and structural manipulation, both to keep tree numbers and fuel loads in check and to adjust species composition to favor those species best adapted to a particular site. Higher incidence of disease and insect outbreaks are now playing a greater role in forest mortality. Prescribed fire is increasingly being used as a positive management tool to maintain existing forests, such as longleaf pine stands, and to restore forests. Reintroducing fire into forest ecosystems is a controversial issue in forest regions as fears of out-of-control fires remain. Homeowners in forest areas and even plantation owners have opposed the use of pre-scribed burning because the costs of mistakes are high. Air quality stan-dards also restrict the ability of forest managers to expand prescribed burning. In addition, timber harvest can play a role: silviculture can mimic fire by thinning overcrowded stands, selecting the healthiest indi-vidual trees to remain, and removing excess trees to reestablish stocking and stand structures that are more fire resilient. Thinning of young stands and prescriptive burning may prove to be effective in reducing the risk of catastrophic fire while reintroducing an important ecological ele-ment into the forest ecosystem.

Invasives

Invasive species that are not native to the United States are increasingly destructive to our forest ecosystems. While nonnative organisms are cen-tral to the economies and cultures of the world—nine crops provide more than 70% of the world's food supply, and 85% of the world's forestry plan-tations (mostly outside the United States) are planted with exotic species—the introduction and spread of invasive nonnative species pose a threat to biodiversity second only to habitat loss and fragmentation (Ewel et al. 1999). Invasives are estimated to account for 42% of vertebrate extinctions worldwide and threaten 35% of species listed under the ESA (U.S. Dept. of the Interior 1998). To elevate the federal response to these threats, Pres-ident Clinton created the National Invasive Species Council in February 2000 to marshal greater resources to control invasive species.

Only 5 to 10% of introduced species become established, and only 2 to 3% can expand their ranges. There are 6,271 identified species of established, self-sustaining nonnatives in the United States (USGS 1998). Expanding global trade, with international movement of raw wood and wooden packaging materials as well as other products, has made the inad-vertent introduction of new forest pests inevitable. The scale and fre-quency of human disturbance to ecosystems give ample sites for colo-nization, allowing for more rapid dispersal and establishment (Ewel et al. 1999). Fragmentation, extensive road networks, and certain silvicultural practices make private U.S. forests more vulnerable to invasive exotics. Invasive species can dramatically alter ecosystem structures and species relationships, changing dynamics that can cascade throughout a system and harm native species.

The American chestnut blight is perhaps the most famous example of the impact of an invasive species on a treasured native tree. At the time that the Asian fungal disease broke out in New York City in the early 1900s, the American chestnut was a major component of eastern deciduous forests, comprising up to 40% of the overstory in stands from Maine to Georgia to Illinois. The blight has infested almost all American chestnuts on the continent, killing the mature trees and allowing young trees to regenerate to only a limited degree before withering (U.S. Dept. of the Interior 1998). Other introduced diseases are further changing forest composition in the East by killing many elms and, more recently, beech. Eastern dogwood is under attack by anthracnose, eliminating this classic understory tree from portions of its range.

Nonnative aggressive plant species that overtake sites and threaten native forests include kudzu (Southeast), various brooms (Pacific), mile-a-minute (Northeast), prickly vine (Oregon), and English ivy (many regions). St. John's wort, a species that has gained fame as an herbal antidepressant, is an invasive Eurasian weed that aggressively displaces native plants, interfering with soil chemistry and nutrient cycling (Westbrooks 1998). Will the commercialization of this special forest product help in controlling it, or will its marketing be an incentive for expansion?

About two hundred introduced insects are considered pests and seventeen are highly invasive, including the gypsy moth and the balsam woolly adelgid. More than 60% of 165 million acres of forest in the Northeast have been seriously damaged by introduced insects. The estimated value of the defoliation and other damage caused by these pests to timber was $2.1 billion per year (USDA 1999). The Asian long-horned beetle, believed to have been introduced in 1996 from China via shipping pallets, has been detected in fourteen states. It feeds on one hundred species of hardwoods, putting 279 million acres at risk. An Asian gypsy moth was also discovered in Oregon and Washington during the 1990s, leading to a $20 million effort to eradicate it.

Air Pollution
Another serious forest health concern is that of airborne chemicals that alter ecosystem dynamics. Although air quality in rural areas has improved over the last decade, in some areas of the East, for example, the

annual deposition of nitrogen is at least ten times greater than what was common a century ago. This nutritional change obviously affects growth and species composition, since some species are better adapted than others to utilize extra nutrients.

An associated concern has been that of acid rain, acid compounds formed by airborne nitrogen and sulphur from fossil fuel combustion and then deposited on vegetation and soils as wet or dry precipitation. Although sulphur concentrations in rainfall are being reduced, chronic nitrogen additions may continue, leading to nitrogen saturation of soils and increased cation leaching (NRC 1998). Intensive studies in the 1980s failed to find direct impacts on forest health over large areas; however, regional impacts on trees—particularly those at high elevations in the East—have been documented (Park 1987). Acid rain damages the leaves and needles of trees, affecting their metabolism and leaching calcium and magnesium. Nutrient replacement through mineral decomposition is unlikely to be sufficient to reverse the losses and restore ecosystem productivity in the near future (Likens et al. 1996).

Urban air pollution, such as ozone generated by automobiles, can damage forests that are relatively distant but in the same airshed. The ponderosa pine forests in the San Bernadino Mountains near Los Angeles, as well as those of the southern Sierra Nevada in California near the growing cities of Fresno and Bakersfield, have extensive ozone dieback. As sprawl and traffic grow, more forests in urbanizing regions are likely to experience ozone damage. Even chronic low-level ozone exposure in rural areas may be hurting forest productivity in many areas of the United States (NRC 1998).

Fish and Wildlife Habitat

As Flather et al. point out in *Wildlife Resource Trends in the United States* (1999), "The distribution and abundance of wildlife is fundamentally affected by landscape structure, which is, in turn, affected by vegetation cover and the manner in which land is used by humans." The changing composition, structure, successional stages, and integrity of private forest resources described earlier in this section all affect habitat quality for forest species.

As forests comprise about a third of the United States, they are exceeded only slightly by rangeland in extent as a cover type (U.S. Dept of Agriculture 1999). Given their extent and diversity, forests provide habitat for a tremendous number of wild species. For instance, U.S. forests are estimated to provide habitat for at least part of the life cycles of 90% of the nation's amphibian, bird, and fish species and 80% of mammal and reptile species (U.S. Dept. of the Interior 1998). Forests and their residents—whether understory plants, insects, rodents, or birds—form an interdependent web of functions and relationships. The many forest types and species, and the complexity of forest relationships, make it very difficult to succinctly describe the fish and wildlife habitat provided by U.S. private forests.

As forests change, so does their wildlife. Forest species vary with successional age. When forest change is dramatic or extensive enough, the very existence of a species may be locally, regionally, or nationally threatened. Forest change can also encourage growth in some species' populations, such as through forest management that favors their habitats. In addition to forest management influences, hunting, increasing urbanization, and other impacts can influence forest species' abundance and health. A variety of interacting factors can synergistically fuel further habitat change.

For instance, as predators have diminished in numbers, herbivores such as deer have dramatically increased in numbers. In areas of the East, where there is an abundance of early successional and/or edge forest habitat, deer populations are at unprecedented levels. Increases in deer populations are further inhibiting the regeneration of cutover areas through browse damage. This in turn is diminishing a songbird habitat (Mac et al. 1998). The overabundance of white-tailed deer has become critical in some regions and "will likely represent one of the more important wildlife management problems during the next decade" (Flather et al. 1999).

As eastern forests have matured over the course of the twentieth century, recovering from their earlier depletion, populations of cottontails and bobcats, associated with early successional forests, have declined. In the Pacific Northwest, conversely, old-growth-dependent species such as the northern spotted owl and marbled murrelet have made their way onto the federal list of threatened and endangered species, while early successional species are on the increase (Mac et al. 1998). Overall populations of forest species are decreasing, while those favoring open and edge habitats are increasing (Rosenberg and Raphael 1986).

Forest Habitat Fragmentation
"Perhaps the most critical problem facing forest wildlife, worldwide, is the systematic shrinking and fragmentation of their habitat," Rosenberg

and Raphael commented (1986), echoing the assessment of many researchers. Fragmentation includes reduction in overall habitat extent, decreases in habitat patch size, and isolation among patches. Fragmentation results from both human and natural causes, including timber harvest patterns, road building, and nonforest development, as well as fire, wind, and flooding. As the USDA Forests Service notes, "A significant difference between human- and natural-caused fragmentation is that human actions are often more frequent, less random, and more permanent than the natural process" (Haynes 2001).

As forests change, so too does habitat suitability for different species. Fragmented forests may not lose species richness overall, but the suite of species will change, often favoring those that are more opportunistic, prefer forest edge to interior, are very mobile, or have very small-scale habitats. In addition to interior species, wide-ranging mammals and species with poor dispersal abilities are hurt by habitat fragmentation. For instance, road development of 1 mile per 1 square mile of forest can reduce effective elk habitat by 40% (Mladenoff and Pastor 1993).

Researchers such as Saunders et al. explain that as patch size is reduced, ecosystem dynamics become driven more by external than by internal forces (1991). When islands of native vegetation become surrounded by a matrix of agriculture, clear-cuts, and/or development, the dynamics of solar radiation, wind, water, and nutrient cycling change across the landscape. Impacts of external forces increase as the ratio of edge to interior area increases. Scientists have estimated that these edge effects penetrate into the forest for a distance of up to three tree lengths. At a certain size and shape, the patch ecosystem becomes overwhelmed by the more developed matrix. Conversely, small openings within larger, relatively intact forests do not generate edge effects, as the forest ecosystem is adapted to the dynamics of relatively frequent, small-scale disturbances (Noss and Cooperrider 1994).

The remnants of forest ecosystems that are considered endangered because of considerable loss of extent, such as longleaf pine forests or old-growth western conifer forests, are in general highly fragmented. For instance, 75 to 90% of late seral and old-growth forests east of the Cascades in the Pacific Northwest were in patches of less than 100 acres (NRC 1998).

Birds have been found to be useful indicators of broad-scale habitat change. Studies have shown that fragmented forests both reduce the numbers of some forest breeding birds and destabilize bird community structure through local extinction and turnover (Boulinier et al. 1998). Severe population declines of breeding birds have been identified in numerous studies, particularly of neotropical migrants (NRC 1998). A review of twenty-seven years of data (1968 to 1994) from the Breeding Bird Survey indicated that of bird species showing significant population changes, those with declining populations outnumbered those with increasing populations. The data indicate that the species with declining populations tend to be associated with older forest stands or with riparian habitat, and the species with increasing populations with younger forest stands (Sharp 1996). Forest interior birds reproduce less successfully and are subject to greater predation and brood parasitism in fragmented forests, leading to declining populations and local extirpation (NRC 1998). Isolated islands of forests have been observed to function as population sinks, drawing birds into areas where their reproductive success will be relatively low.

Although data vary with the particular species, in general neotropical bird migrants tend to be "more abundant in landscapes with greater proportion of forest and wetland habitats, fewer edge habitats, larger forest patches and forest habitats well dispersed throughout the landscape" (Flather et al. 1999). In addition, the diversity and abundance of neotropical migrants decrease as area housing density increases, regardless of forest patch size. Therefore, some researchers believe that the benefits of forest regrowth in regions such as New England are negated, from a habitat perspective, by urban sprawl (Friesen et al. 1995).

In the Pacific Northwest, "where the depletion and fragmentation of commercially harvestable Douglas-fir forests rivals that in any other forest type on earth," very extensive areas (more than 50%) of mature and old-growth forest have been harvested in the last thirty years (Rosenberg and Raphael 1986). This has rapidly converted continuous forests into a mosaic of patches in successional stages from brush to remnant mature stands, characterized by abrupt edges. Severe edge effects have been observed in such areas (Sharp 1996).

Threatened and Endangered Species

In the USDA Forest Service publication *Biological Diversity: Status and Trends in the United States,* the authors note that "biotic integrity can be inferred by looking at the percentage of species considered threatened by

Table 2-4.

Forest type associations among 667 threatened and endangered species listed under the ESA as of August 31, 1992

(Species may be listed under more than one forest type.)

	Deciduous	Evergreen	Mixed	Total Forest	% of Total T&E
All T&E	128	178	110	312	47%
Plant	39	60	34	109	16%
Animal	89	118	76	203	30%
Mammal	11	21	15	29	4%
Bird	7	41	16	48	7%
Reptile	7	10	11	16	2%
Amphibian	2	2	4	6	1%
Fish	28	28	19	54	8%
Snail	8	2	2	10	1%
Clam	19	8	7	25	4%
Crustacean	6	1	2	9	1%
Insect	1	5	0	6	1%

Source: Flather et al. 1994

extinction" (Langer and Flather 1994). According to a World Resources Institute report they cite, 4% of U.S. birds and mammals and 12% of plants are threatened. The degree of threat to U.S. animals is similar to that of other temperate and tropical countries, and is higher for plants. As of September 30, 1999, 1,213 species were listed under the federal ESA as either endangered or threatened, with an additional 256 candidate species. As table 2-4 illustrates, forests provide habitat for approximately half of listed species. More threatened and endangered (T&E) species are listed for forests than for other land use types. The habitats of almost 80% of T&E species are on private lands, a significant amount of which is forested (U.S. GAO 1994).

According to another USDA Forest Service report, *Species Endangerment Patterns in the United States,* the primary cause of endangerment of forest species is habitat loss; while threats from introduced species constitute the second major cause (Flather et al. 1994). Based on similar species and environmental attributes these researchers identified ten "high endangerment" regions in the United States (see table 2-5). Old-

Table 2-5.
High species endangerment regions in the United States

Region	Forest-associated T&E Species
Southern Appalachia	64%
Peninsular Florida	64%
Eastern Gulf Coast	63%
Southern desertic basins, plains, mountains	22%
Arizona Basin	44%
Colorado/Green River Plateaus	34.5%
Central desertic basins and plateaus	33%
Southern Nevada/Sonoran Basin	17%
Central/Southern California	14%
Northern California	37.5%

Source: Flather et al. 1994

growth, riparian, and savanna forests, as well as freshwater ecosystems,[2] are at-risk communities with high numbers of threatened and endangered species. For example, the longleaf pine forest ecosystem, a highly endangered forest type, is home to twenty-seven federally listed and ninety-nine candidate species (NRC 1998). If all of the candidate species occurring in California were listed, this state would be shown to support the greatest concentration of threatened and endangered species in the lower forty-eight states. Due in part to increasing urbanization and rapid changes in forest composition in the last forty years, the Pacific Northwest is emerging as a new region of species endangerment. The forests of the Cascades are a notably diverse region because of their heterogeneity; they support three times more mammals and two times more breeding birds than the coniferous forests of the southeastern coastal plain (Harris 1984).

2. Nearly half of all U.S. mussels are listed or proposed for listing under the ESA. Of the remaining mussel species, most are imperiled as well. The American Fisheries Society listed 364 North American fish taxa as either threatened (114), endangered (103), or of special concern (147). Two hundred fourteen native Pacific salmon stocks (23% of all Pacific salmon stocks) were listed as at risk in California, Oregon, Washington, and Idaho (Langer and Flather 1994).

Reasons for species endangerment vary by region: urban development, for instance, is the primary threat identified for Florida, the Gulf Coast, and Southern California. Florida species are also threatened by forest clearing, agriculture, and fire suppression. Additional threats in the Gulf Coast include shoreline development and "human-caused mortality." Agricultural development and aquatic contaminants, including sedimentation, affect threatened mollusks in southern Appalachia. In Northern California, agricultural development and impacts from heavy equipment and urban development are priority threats. Species in the West are primarily threatened by impacts of grazing, water development, introduced species, mining, and drilling. Recent listings of many salmonid populations in California, Oregon, and Washington are attributed to many factors related to land use change that has reduced spawning and rearing habitat. These factors include destruction of riparian vegetation, removal of large woody debris from streams, and increased sedimentation from roads, timber harvest, agriculture, and other development. Dams have also played key roles in reducing salmon habitat and limiting passage in certain major watersheds. In addition, salmonids have been threatened by overfishing and other factors unrelated to land use.

The broad impact of species listings under the ESA can be seen through the dramatic increase in HCPs submitted by forest landowners. Under Section 10 of the ESA, nonfederal landowners can prepare an HCP as part of an application for an incidental take permit. The HCP seeks to minimize and mitigate the impacts of the approved management activities, allowing for a limited or incidental degree of harm to species in return for the landowner's commitment to maintain or enhance habitat overall in accordance with the plan. As of December 1999, the federal government had approved 266 HCPs covering about 20 million acres, with more than 150 in development. An analysis of the U.S. Fish and Wildlife Service's database of HCPs prepared for this book showed thirty HCPs covering 8. 5 million acres of forestland,[3] with individual plan areas ranging from 12 acres to 5 million acres (International Paper). Therefore, HCPs related to forest species account for about 42% of the total acreage subject to incidental take permits to date. Although

---

3. An additional plan, for Crown Pacific's Hamilton Tree Farm in Washington, is listed with no acreage indicated.

there is intense controversy over the effectiveness of HCPs in maintaining populations of threatened and endangered species on managed private lands, the sheer number of acres under HCPs is an indicator of the degree to which land management practices on private lands can affect the survival of species.

Water Resources

Water provision is an immensely valuable function of forest ecosystems. Approximately 60% of the nation's total stream flow—the primary source of water supplies—comes from forests. Forested watersheds catch and filter water, as well as regulate flow and moderate flooding. The nature of the forest and the course of forest management in watersheds can positively or negatively affect the supply and quality of water produced and thus all of its beneficial uses, including drinking water, irrigation, livestock, power production, recreation, and habitat. Forested waters and upstream habitats can be influenced not only by local impacts but also by downstream activities, such as amounts of water withdrawal.

With population growth and intensification of land use, water demand is rising and good-quality, unpolluted supplies are increasingly scarce. Some experts suggest that current levels of freshwater withdrawal already exceed available, dependable supplies (U.S. Dept. of the Interior 1998).

As illustrated in the previous section, changes in instream habitats from a variety of uses have led to the ESA listing of more than one hundred species of freshwater fishes and placed more than 250 species in danger. In 1994, 56% of endangered species and 58% of threatened species were associated with aquatic and riparian habitats. Damming, channelization of rivers, urbanization, agricultural development, road building, logging, water pollution, and the introduction of exotic species have all played a role in the loss of aquatic species. "The total effect of these developments is the alteration of stream ecology as evidenced by changes in the migration patterns of fishes, in stream water temperature and nutrient levels, in water chemistry, and in biological diversity" (U.S. Dept. of the Interior 1998). According to the USDA Forest Service report *An Analysis of the Water Situation in the United States: 1989–2040,* only 28% of surveyed watersheds in the United States were in prime condition, while 50% required special soil and vegetation management and 22% required restoration (Guldin 1989).

While the USDA Forest Service considers forest water quality to be "good" overall, watershed functioning can be negatively affected by removal of tree cover and vegetation in both riparian and upslope areas as well as by other forest management activities. Nitrate levels can increase owing to timber harvest or fertilization (Brown and Binkley 1994). However, the major water quality problems influenced by forest practices fall under the category of nonpoint-source pollution: increased sedimentation and instream temperatures. Both are critical issues for effectiveness of instream habitats. According to the 1996 National Water Quality Inventory, nonpoint-source pollution is the nation's largest water quality problem, affecting some 40% of surveyed water bodies. In California, for instance, almost every watershed in the redwood region (coastal watersheds from Monterey County to the Oregon border) is rated as Priority Category 1 under the EPA's Unified Watershed Assessment because of sedimentation and/or temperature impairment from sources that include silviculture (EPA 1998).

Sedimentation can change instream structure by adding more fine particles to gravel beds, increasing erosion of stream banks, reducing stream depth, widening streams, and filling pools (Brown and Binkley 1994). According to the 1982 NRI, forestland contributed 10% of total sediment discharge from nonfederal land (NRC 1998). Sedimentation is exacerbated by the heavy impact of rain on exposed forest soils and roads; compaction of soil, which reduces its infiltration and increases runoff; and the growing likelihood of mass wastings from timber harvest and road-building activities. In research at the H.J. Andrews Experimental Forest in Oregon, clear-cut slopes were found to account for four times and roaded areas for fifty-one times as much annual debris flow as forested slopes (Brown and Binkley 1994). The EPA estimates that road construction and use contribute up to 90% of forestry-related nonpoint-source pollution (1997). The likelihood of sedimentation varies in watersheds depending on soil type, slope, climate, and other factors (Brown and Binkley 1994). According to a discussion in Guldin (1989), undisturbed mature forests generate very low annual sediment loads (less than .5 tons/acre). Well-managed timber operations may generate an additional 1 ton/acre. However, many logging operations generate 10 to 15 tons/acre. Intensive site preparation that seeks to minimize slash and maximize bare mineral soil can generate as much as 100 tons/acre.

Forest practices that change instream temperature more than about 2°C from natural temperatures may alter the development and success of fish populations; some species, such as coho salmon, are more sensitive than others. Removal of forest canopies over streams can raise stream temperatures by 5°C or more. However, areas left with sufficient riparian canopy as a buffer can keep instream temperature change below the 2°C threshold (Brown and Binkley 1994).

Forest watersheds can be effectively managed to improve water quality and flow by protecting and enhancing riparian vegetation for filtration, cooling instream temperatures, providing wildlife cover and instream channel debris, and modulating runoff. The condition and management of upslope areas can also benefit water resources: well-designed and maintained roads and timber-hauling systems reduce sediment inputs without destabilizing slopes, and application of chemical fertilizers or herbicides can be well timed to minimize runoff. Even with such excellent practices, however, it can often take many decades to ameliorate the impact of past activities.

Since the passage of the Clean Water Act, states have been encouraged by the EPA to implement best management practices (BMPs) to control negative impacts to water quality. As of 1992, twenty-three states had voluntary and thirteen regulatory BMP programs; five used a combination. When state programs included compliance monitoring, 85 to 90% reported compliance. In states with voluntary programs, compliance fell to less than 50%. NIPF and smaller landowners are somewhat less compliant than IPF and larger landowners (Brown and Binkley 1994).

Forest Carbon Stores

One of the key ecosystem functions forests provide is sequestering carbon dioxide (CO_2): absorbing CO_2 through photosynthesis and storing it as carbon in their biomass. The increasing level of atmospheric CO_2 is the major contributor to global warming and climate change. Forests can be sinks or sources of CO_2, depending on their management. They are potentially the largest and most expandable long-term carbon sinks in the world; therefore they are very important to any strategy to ameliorate global warming. However, they are also the second-largest source of anthropogenic CO_2 emissions, owing to forest loss and management impacts. Forests that are frequently and highly disturbed become sources, releasing more carbon than they have stored.

Subject to historic and continuing loss of area, decreasing age, and increasing management intensity, private U.S. forests have declined as carbon sinks. In 1997, U.S. forests as a whole offset only 5.9% of U.S. emissions, storing a net of 89 million metric tons (MMT) against emissions of 1.5

Figure 2-11.
U.S. forest carbon stores are declining

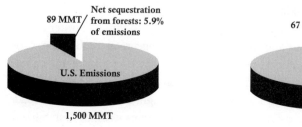

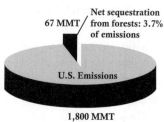

billion metric tons (BMT) of CO_2. This downward trend is expected to continue to 2010 (figure 2-11) and beyond, if current forces remain unaltered (U.S. Department of Energy 1997). Further, the 2nd U.S. National Communication to the UNFCCC projected that forest sequestration would decrease by 25% between 1990 and 2020. As this figure is from Birdsey and Heath's 1995 report, it is likely to be conservative, as their data do not reflect more recent forest changes, such as increased timber harvests in private forests. However, if current trends were altered, as described below—by preventing forest loss, reforesting former forestlands, and growing older forests—net forest sequestration could increase to offset at least 20% of U.S. emissions.

Global recognition of the need to slow or even reverse the increase of atmospheric carbon dioxide releases offers a great opportunity for management of U.S. forests, particularly private forests, to rebuild our domestic carbon banks. This can be done by growing older forests, which store more carbon; protecting existing forest carbon reservoirs from conversion; replanting understocked forests; and restocking former forest soils. Sustainable harvest of wood products as building materials and other long-lasting uses will contribute to an improved carbon budget as well. As nations around the world improve cooperation in reducing global warming gases, a new market may emerge for the storage of carbon by forests. Although nascent and dependent on the evolution of international policy, this market could provide financial returns for private forest conservation in the United States, as described further in chapter 3.

Forest carbon is stored as biomass accumulated on the outer layer of cambial tissue, increasing as trees grow taller and larger in diameter with age. This biomass continues to accumulate for the lifetime of the tree, even after it reaches five hundred years of age or more. Biomass, and therefore forest carbon, also accumulates in the soil, with greater stores in older, less disturbed soils and forest floors. Older trees, and therefore forests, store more carbon in total *amounts* both over time and annually than do younger trees and forests, although younger forests may sequester carbon dioxide at faster *rates*.

In general, forests accumulate increasing amounts of carbon annually until forest stands reach the culmination of annual increment, at which point annual increases plateau and slowly decline. Nonetheless, as noted above, recent research demonstrates that increases continue into hundreds of years. With normal mortality, the carbon stored in tree biomass (roots, boles, branches, bark, leaves) decays slowly, becoming primarily soil carbon, with minor loss, less than 5% of total forest carbon, to the atmosphere.

Forest soil carbon is stored in two forms, labile and stable. Labile carbon is found in the upper soil layer, and can be released when disturbed. It accumulates from humus and leaf litter decay, and can vary with growth and release from disturbance in short time frames (years to decades). Stable soil carbon is generally very difficult to volatilize and release, and varies very little over long periods of time (thousands of years).

With disturbance events, more carbon may be released to the atmosphere, depending on the type of disturbance. Disturbance events include, in order of increasing disturbance and carbon release, major pest and disease outbreaks, wind and ice storms, fire, harvest, and conversion. Carbon stores are generally not lost to disease or pest outbreaks, but future gains are diminished or halted. Wind and ice storms produce a similar pattern. With fire, however, increased amounts (up to 10%) of on-site labile carbon are vaporized and released into the atmosphere.

Typically 40 to 50% of on-site forest carbon is released through logging within the first five years after harvest, owing to increased soil disturbance, exposure, decay, and site preparation activities. Some forest carbon is removed from sites during harvest and processed into timber products, resulting in stores of roughly 15% of forest carbon. This carbon then continues to decay more or less slowly depending on the product type and its treatment. Wood products used within buildings may have a lifetime of hundreds of years, paper products a lifetime of weeks. Typically, less than 50% of forest products last more than five years. With conversion of forests to nonforest areas, up to 100% of on-site carbon stores are lost.

Figure 2-12.
Forest carbon after harvest (of an 80-year-old Douglas-fir stand)

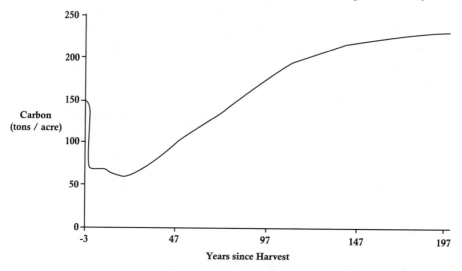

Forest management can also enhance reaccumulation of forest carbon through prompt reforestation, lighter-impact harvest practices, and longer rotations. However, it takes a forest stand about the same number of years to recover carbon after harvest as the number of years of forest age at harvest. Thus it takes at least eighty years to reaccumulate the carbon stored in an eighty-year-old forest (figure 2-12). With shorter rotations, more time is required to restore the carbon balance than is allowed by the rotation length because forest carbon is lost in the process of timber harvest. Thus one ninety-year rotation stores more carbon than three thirty-year rotations, and a greater total volume of timber products is gained.

Changing Forest Conditions and Implications for Biodiversity

Forests have changed considerably since Europeans arrived in North America. But this was not the beginning of change, as human activity, climate, migrations of species, and other natural forces have sculpted forests for millennia. Some forces of change are ongoing; others are episodic. The difference from two centuries or two millennia ago is that impacts originating with human activity are becoming dominant. Forest management, or lack of it; fire, or lack of it; intensifying land use of every sort; demands of growing populations; and unanticipated environmental effects of human activity, such as global warming, are all powerful sources of forest change.

All the forest types of the United States have gone through profound change since the country's settlement. Beginning with New England and gradually sweeping across the continent, forests have been cleared. Many have regrown. Others have been permanently converted to other uses. The forests that have come back are very different than the original forests, with different species composition, structure, and function. These changes are driven by the reasons we have surveyed above: fragmentation, fire suppression, silvicultural favoritism of different species, microclimate alteration, introduction of exotic pests, and loss of other ecosystem elements.

Some forests, such as the hardwood forests of the central Midwest, have proven to be quite resilient, losing few species. Others, such as the vast eastern white pine forests of lower Michigan and northern Wisconsin, were replaced by hardwoods after huge slash fires in the mid-1800s burned the clear-cut ground and changed the soils. The oak savannas of the Midwest, South, and West have been mostly cleared, usually for agriculture. Where forests have regenerated, in Wisconsin, for instance, aspen and birch have dominated. In New England, early successional sun-loving species such as birch, red maple, and oaks are expanding while long-lived, shade-tolerant trees such as sugar maple and eastern hemlock are declining. Fire suppression in the Southeast has changed forest composition, allowing for denser, woodier understories. Hardwoods have moved into longleaf pine stands where prescribed burning is not practiced. Similarly, southwestern ponderosa pine forests are unnaturally thick with young trees, decreasing stream flows and building up potentially catastrophic fuel loads. On the other hand, aspen, an early successional species, has been decreasing in area. Pinyon-juniper forests essential for maintaining stable watersheds in the arid region are being cleared for pulp and conversion to grasslands. California's redwood forests have come to be populated with a much greater amount of tanoak, a vigorous stump-sprouter encouraged by clear-cutting. The huge trees and massive stores of biomass that characterized the Pacific coastal forests are largely gone, dramatically altering microclimates and habitats across large areas. In general, only some 15% of old-growth forests remain, mostly in Alaska (NRC 1998).

Forest change can either enrich or impoverish biodiversity. Why do we care about biodiversity of forests? One very practical reason is that it maintains options within a system, thus providing greater resilience to disturbance. In a world of change, maintaining options is essential to survival.

Biodiversity is therefore an expression of ecosystem functionality. As noted in *The Status and Trends of the Nation's Biological Resources,* levels of biodiversity depend on the interaction between forces that favor homogeneity of a natural system and those that favor heterogeneity. "Maintaining the heterogeneity on which natural diversity is founded while preventing the extreme homogenizing tendencies of human beings is the key to maintaining . . . natural heritage" (U.S. Dept. of the Interior 1998).

To briefly summarize its elements, biodiversity encompasses all the interactions and processes of an ecosystem. It includes not only the numbers (or richness) of species in an ecosystem, but other elements as well:

Diversity of species: When a large number of different species is present, the chances are greater that key interactions will keep working even if one species is in trouble.

Diversity of structural stages and life cycles: For example, it is important to have some trees in seedling stage, some mature, and some in between. Each provides different kinds of structure and relationships with other organisms. Insects in caterpillar and butterfly stages are as different from each other as insects of separate species. A variety of life stages ensures that a catastrophe killing a population in a crucial life stage will not wipe out an entire species or system.

Diversity of life in functional and nutritional niches: A variety of plants, plant eaters, and predators—dead trees, cavity makers, and cavity users, for example—ensures that there is more than one species capable of carrying out a key function.

Diversity of landscapes: Diversity is important across and within landscapes. All elements of a system cannot be omnipresent; rather they reside in patches varying with land forms, microclimates, and histories of disturbance and renewal.

The kinds of forest changes that have occurred and are occurring in the United States have reduced biodiversity. Overall, as the data have shown, U.S. private forests today are generally younger, simpler, and more homogeneous than their predecessors. Fragmentation is an accelerating phenomenon, reducing forest patch size and isolating key habitats. Forests are characterized more by edges and less by interior habitats. Naturally regenerated second- and third-growth forests often suffer from too little stocking or too much. Many NIPF forests consist of poorly formed and genetically inferior individuals remaining after a history of high grading. Natural forests are increasingly being replaced by simplified plantations. Key habitat structures such as large, limby old trees, cavity trees, standing dead trees, and downed logs (coarse woody debris) are missing in many stands. Vertical and horizontal diversity is often lacking in stand structure. Soil banks of nutrients and organic matter have been drawn down by frequent harvest and regeneration. Soil structure and organisms have been damaged by compaction. As the NRC reported, "Increased management intensity for tree-fiber production creates greater uncertainty, if not actual decline, in the delivery of other natural and societal benefits from forest ecosystems. Forests managed with greater attention to tree growth and harvest removals [are] simpler in terms of structure (spatial heterogeneity) and biodiversity than unmanaged forests" (1998).

On the other hand, although forests have changed dramatically since European settlement, so far they have proven their resilience after two centuries of profound disturbance. Further, the United States is not losing forestland at the rate it once was and forest area overall is relatively stable. Since the turn of the twentieth century approximately 50 to 80 million acres of cutover land that had not regenerated are now mostly reforested. Large, industrial landowners are investing in reforestation and stand improvement, having put behind them a more destructive history. More Americans own forests than ever before. Understanding of biodiversity and the application of ecological science to forest management is expanding. Forest managers are increasingly applying environmental standards to their work, seeking to exceed the minimum requirements of regulation. Conservation of forestland is moving beyond public acquisition to new models of private forest protection that integrate resource protection with resource production. There are many opportunities to rebuild biodiversity and enhance the vitality of private U.S. forests, as we will discuss later.

Chapter Three

Threats to Private Forests and Barriers to Their Conservation

Forest loss is the expression of a complex web of factors including economics, changing markets, demographics, land use history, culture, and natural disturbances. This suite of threats and barriers is what must be addressed to successfully accelerate the conservation of private U.S. forests. In this chapter we will draw on the overview of America's forests laid out in chapter 2 to highlight the major threats to their continued existence and functionality. In addition, we will lay out the primary barriers to forest conservation faced by landowners and all stakeholders.

The Continuum of Loss: Degradation, Fragmentation, and Conversion

Forest conversion does not happen overnight. Usually by the time the forest property is being sold as residential or recreational real estate, a complex of forces has been at play in the vicinity for some time. Outright conversion occurs once the value of the property as forest can no longer compete with its value as some other use, whether that is residential, agricultural, or other development. The balance is only tipped after some years, often decades, of gradual erosion of forest value relative to other values. Once a forest owner or that owner's neighbor has an offer to sell his or her land for five times its forest value, the process of conversion is too far along for any effective intervention. However, complete conversion to nonforest use is simply the easiest kind of forest loss to track. As we have seen in our survey of the status of America's forests, conversion is not black and white, but the terminus of a continuum of loss.

Statistically, acres are still counted as forest even when their ecological and economic functionality as forest may have been dramatically diminished by a history of simplification, fragmentation, and other forms of ecosystem degradation. We can somewhat arbitrarily draw a line at 10 acres as a minimum parcel size below which statistically a forest property has been functionally converted to residential use. However, other forms of forest loss are more subtle. Parcel size, tree stocking, forest composition and structure, introduction of other uses and species, and surrounding uses all influence the degree of conversion risk of a particular forest property.

The factors that influence the degree of conversion risk include those that are endogenous to the property and those that are exogenous. Endogenous influences include the quality of past forest management practices: if they have been poor, the quality of the existing forest and its productivity may be degraded, reducing its relative economic value as forest. Exogenous influences include the growth of other uses in the vicinity of the forest property: if residential or agricultural demands have intensified, and other forest properties in the area have been converted, the risk is increased that other properties will also be converted. The growth of alternative uses is fueled by overall economic and demographic trends. These include general population growth, growth of the age thirty-five to sixty-four segment of the population who are in their prime working/buying period, higher disposable incomes, desire for increased living space, and expansion of second-home purchases.

Fragmentation is a factor that operates both as an internal and an external influence and significantly adds to forest conversion risk. Fragmentation affects the ecological functions of both the fragmented parcel and its neighbors, even if those tracts are sizable and well forested. Therefore, the effects of the increasing rate of fragmentation of medium and larger forest properties into smaller parcels will not be limited to the 2 million or so acres broken up annually, but will radiate out into the landscape. The fragmentation of larger forest tracts into smaller parcels is both an expression of and fuel for further development of other uses in forest areas.

Forest conversion is finally a market-driven phenomenon. Landowners make decisions about their properties based on formal or informal highest-and-best-use analyses, weighing their needs and options to realize financial and nonfinancial value from their properties, whether as forest or other uses. Judgments and decisions made in the past affect their options in the present. Forest conversion therefore happens within a context of markets, landscapes, and historic management practices. Any strategy for forest conservation needs to find the means to intervene along the conversion continuum, and, where possible, get ahead of emerging trends.

Creeping Urbanization Undermines Forest Integrity

The fragmentation of forest lands into smaller parcels is not the only expression of creeping conversion. The forest fabric is frayed and forest integrity undermined by the gradual extension of urban infrastructure—especially improved road networks—into rural, forested areas. Urbanization, or increasing density of people and buildings, sends waves of intensifying development out into the countryside where land is cheaper and regulations often fewer. Land at the edge of urban areas is frequently rezoned and reassessed in anticipation of development. This becomes self-fulfilling as forest owners cannot carry the higher costs, and as alternative use values become economically irresistible. Ultimately, when the urban-influenced real estate market is just over the horizon, what *can* be subdivided in a rural area *is;* and ultimately, what is subdivided *will* be built as the market grows. The more accessible, gently sloped properties go first. The more rugged and remote take longer. As subdivision and subsequent building expand in a forest area, the forest ecosystem is further impaired by more clearing, more road building, expansion of utilities, altered hydrology, soil compaction, and introduction of exotics. This process may take five years or thirty years to run its course, depending on economic cycles. Meanwhile, the urban wave continues to ripple out, pushing the leading edge of its influence into previously remote areas.

An associated threat is that as forest regions urbanize, the loss of available timber drives mill closures. The reduction in markets for different kinds of harvested timber—or any timber—restricts the options of remaining forest owners in their forest management. This can undermine the financial attractiveness of forest ownership and contribute to sale of forest properties for residential or other uses.

Intensification of Forest Uses and Conflicts

We are faced with the twin, related phenomena of increasing population and increasing demands on forests. More people require more land to support them: University of Georgia ecologist Eugene Odum calculated that on average it took 5 acres of land to supply the needs of one American, including 1 acre for development, 1.5 for fiber, 1.5 for food, and 2 for other ecosystem services (1998). More people want more out of forests: wood for the booming building market, beautiful settings for new (often second) homes, recreational opportunities for greater leisure time, habitat for threatened creatures, increased carbon stores to reduce global warming gases, and clean water for all uses. Forest use is intensifying, whether as plantations for fiber or as rural residential subdivisions. At the same time, forest acres lost to development are growing—perhaps faster than at any time in the last thirty years or more. Some forest resources, such as mature timber, are in tight supply. Increased population density in forest regions has led not only to less available timber for harvest, but also to the negative impacts to ecosystem values that development and fragmentation bring. Increasing but varying demands on a limited resource lead to conflicts over uses, setting different forest stakeholders against each other. The result calls for greater protection of private property rights as well as for greater protection of public benefits.

People and Forests Have Different Time Frames

People are the primary force shaping forests today, both through direct impacts such as timber harvest choices and indirect ones such as climate change and the introduction of invasive, nonnative species. As a species we are challenged in the management of systems such as natural forests that have life cycles much longer than our own: while our life expectancy is about eighty years, most tree species of that age are just hitting their biological stride. Our financial needs tend to drive timber harvest more than silviculture and the biological requirements of the ecosystem. The challenging time frame of forest management is worsened by accounting norms that value short-term cash over longer-term asset appreciation.

Investment analysts who consider seven years a distant investment horizon heavily discount the returns of intensive plantation forest management rotations of twenty to twenty-eight years. How can older, more complex forests—with their unique habitat values, carbon storage capacity, and high quality timber—ever be reestablished on more of the private forest landscape when peoples' financial and biological horizons are so short by comparison?

Lack of Continuity in Forest Ownership and Stewardship

The relatively short period of people's lives leads to a lack of continuity in forest ownership and therefore in forest stewardship. As we discussed earlier, only 10% of owners, representing 30% of private forestland, have held their property forty-five years or more. Each new owner inherits the results of the decisions of past owners, expressed in the forest condition and degree of building that has occurred on the property. This history restricts the management options of succeeding owners. Each new owner, whether industrial or nonindustrial, usually has different goals for the forest to meet. Forest management plans made by previous owners, if any, are often not implemented by new owners but are superseded by new decisions based on the new owner's financial or other requirements. Therefore the forest ecosystem, a long-term asset, is constantly asked to respond to new, short-term needs. Lack of continuity in stewardship shortchanges the ecosystem, limiting the beneficial impacts of previous forest investments. The result limits the future options of landowners and society. Frequent turnover tends, as well, to increase fragmentation of forestland. Most parcelization occurs in the process of an ownership disposition as either the seller or the buyer seeks to capture any real estate value that may have accrued to the forest property. Therefore, the current increased disposition of both industrial and nonindustrial forestland further threatens forest integrity.

The continuity of stewardship of family-owned forests is especially threatened by a suite of succession issues. With a quarter or more of private forests owned by people who are at least sixty-five years old, stewardship succession is a growing concern. Family forestlands can be subject to unplanned or excessive timber harvest, and to being broken up, if the family has not prepared well for transfer of ownership from the older to the younger generation. Some families are astonished to discover the tremendous increase in asset value of their land and timber, and therefore may have failed to make adequate preparations for funding estate taxes. Heirs are increasingly distant from the property, and may be disinterested in the demands of forest ownership and stewardship. They may prefer to cash out, especially if the timber or real estate value has increased considerably. Even if some family members want to maintain the forest property and continue the stewardship of the older generation, the form of family ownership may make intergenerational transfer difficult. Ownership forms such as undivided interests or family partnerships can paralyze decision making as family interests multiply and ownership fractionalizes further with each succeeding generation.

Industrial ownerships, such as publicly traded entities, may not have limited lifetimes the way people do. But that does not help sustain forest stewardship any better over the long term. Forest management goals change based on a variety of company objectives, from changing markets, products, and strategic positioning to changing personnel. Further, industrial owners review and turn over their portfolios of forest properties as their financial goals require, continually seeking to enhance shareholder returns. Prime redwood forestlands in Mendocino County, California, for instance have typically been under at least three industrial ownerships in the past fifty years. Are there new forms of ownership that can provide greater continuity of stewardship, helping sustain forests on their own time frames?

Lack of Investment Capital and Instruments That Support Stewardship

Compared with the long-term nature of forests as assets, the available financing instruments have too short a time frame. The time value of money is a significant barrier to forest conservation and stewardship because long-term yields are undervalued. Natural diversity is also undervalued since financial returns on private forests are primarily generated through management and harvest of fast-growing commercial tree species and through real estate spin-offs. Forest stewardship is undermined by needs for short-term liquidity, whether to pay debts or distributions to investors or to fund unforeseen medical or other emergencies. In fact the funding of financial emergencies was highlighted as the top barrier to forest conservation in the NIPF survey conducted for this book by Mater

Engineering. Forest stewardship by individuals is further hampered by lack of financing alternatives for up-front forest improvement investments that may not provide a financial return until a distant timber harvest. Existing government cost-share programs are hampered by poor funding and sometimes complex requirements that limit landowner involvement. Investments in ecosystem restoration are inhibited since they do not yield financial returns and current tax policies do not allow for their expensing by many smaller owners. Are there different forms of working capital and investment equity that are better aligned with the characteristics of forest ecosystems?

The needs of highly competitive financial markets have driven the depletion of natural forest capital. Even the most committed private forest landowners faced capital constraints on their stewardship as they vied for investment during the 1990s in a world of Internet start-ups and seemingly endless 20%-plus stock market index returns. Fundamentally we need to address the question, What is the sustainable rate of return from forests? Is this rate of return sufficient to attract ongoing investment?

People's Perceptions of Forestry and Its Impact on Forest Health

The ecological values and health of nonindustrial private forests, in particular, are threatened by a combination of either too much management or too little. That is, many have been logged poorly in the past and either are continuing to be overlogged or are being neglected. One problem is that forest management all too often is equated by landowners, foresters, and others with logging. This leads people to pay attention to the forest when it has timber value, but not otherwise. It is important to remember that timber harvest pays the bills for the services of consulting foresters and contract loggers. Foresters and loggers are frequently paid a percentage of the harvest value, or on the basis of volume harvested. This creates a natural bias toward commercial logging over other forms of forest management.

The public and environmental organizations often think of forest management as logging too. But extensive exposure to poor timber harvest on public and private forests has turned public opinion against logging in particular and forestry in general. In addition, too often landowners have seen or heard of loggers leaving land in bad shape and even cheating their clients. The result is increasing distrust of loggers, foresters, and state forestry personnel by small and even medium forest landowners, environmentalists, and the general public. This creates a huge hurdle for good, stewardship-oriented forestry to overcome as well, as it has been "painted with the same brush" as the poor variety. Many individual landowners therefore believe that the best stewardship may be to leave the forest alone rather than risk degrading it further.

Other forest owners may simply lack the necessary information or time to be more engaged with their forests, given that their forest property is usually an adjunct to other land uses or not central to their livelihood. Some landowners who do log often do so without a clear stewardship objective, but simply want to take advantage of a hot stumpage market or to provide cash for family needs. In any case, most small and medium forest landowners will only rarely harvest timber. Because forest management information is only occasionally relevant to them, they are not motivated to seek it out. When the time comes when such information may be relevant, they usually do not know where to find it. Sampson and DeCoster call this the "catch-22 of occasional relevance."

The increasing number of small landowners makes the challenge of engaging them in the stewardship of their forests more difficult than ever. As NIPF lands are carved up into smaller, more residential holdings, careful timber harvests and other forest stewardship activities to benefit wildlife or water quality may occur less and less, hurting both environmental values and timber supplies.

Forest Stewardship Is Neither Cheap Nor Easy

Forest landowners who want to steward their forests and make investments in forest health and ecosystem restoration are often stymied by lack of accessible, practical information in language they can understand. Data on the economic and ecological benefits of natural forest management, incorporating the stewardship of both timber and nontimber resources, are still not widely available. Further, stewardship-oriented forestry is scientifically grounded and requires a greater degree of technical expertise than is available to many forest owners. Forest stewardship calls on various sciences, including not only silviculture but also wildlife biology, hydrology, geomorphology, soils, and others. Larger landowners are more able than smaller ones to assemble specialists.

Often there are no comprehensive, accessible sources of stewardship assistance for a landowner to call on. State and federal agencies have different mandates, jurisdictions, and structures that have kept them from providing easy, centralized help for the multiresource stewardship issues that forest owners face. The array of landowner assistance programs offered by government agencies can be confusing and poorly communicated. Further, even popular and effective programs such as the Stewardship Incentives Program (SIP), in demand by landowners, have been defunded by Congress. Most government programs to assist private forest landowners were developed to address single issues, such as timber production, wildlife management, or water quality, with different agencies assigned jurisdiction. This severely hampers the coordination of program implementation and limits opportunities for increased efficiency.

In addition to extension foresters, consulting foresters and loggers are typically the primary sources of information and services for individual forest landowners. These service providers are in short supply in many forest areas. Worse, they may not be well regarded and may not be current with advances in ecological management. Unfortunately, many embrace a traditional growth-and-yield, get-the-cut-out culture and therefore can be unwilling or unable to genuinely address a landowner's stewardship goals. Conservation organizations, including land trusts and wildlife groups, offer some assistance to forest owners as well. However, their focus and resources can be limited. They too can suffer from poor reputations among larger forest owners if they are seen simply as preservationists and tree huggers out to "lock up my property."

Forest stewardship costs money. Many landowners lack the cash to invest in forest stewardship or ecosystem restoration activities if they are not undertaken within the context of a timber harvest that generates a profit. Given the limits on attention, time, and cash of most individual forest owners, is it surprising they do not make more investments in forest stewardship?

Shifting Markets Shape the Forest

Complex global markets are increasingly forceful in shaping private U.S. forests. As we have discussed, landowners' financial needs are key drivers of forest management decisions. Those needs operate within the supply-and-demand flows of a globalized wood basket. The cascade of economic crises in 1998 known as the "Asian flu" dramatically changed the fortunes of Asia's emerging "tiger" economies, such as Thailand, Indonesia, and Korea, and also wreaked havoc with U.S. wood exporters to the Pacific Rim. High-quality export log prices plummeted and supplies formerly bound for foreign markets stayed in the United States, pulling down prices for most Pacific Northwest species. Pulp markets have become completely international, with producers sourcing chips and pulp routinely from around the world. U.S. forests and their stewardship are affected by this web of commerce.

Shifting markets challenge forest stewardship because each timber harvest decision today affects the kinds of forest products that may be available in twenty or fifty years. As species and log grades have become scarce, demand has shifted. Species considered low grade in the past are the focus of harvest today. In addition, scarcity has led to new technologies, with new engineered wood product facilities or small mobile chip mills driving increased harvest of previously undesired species in previously remote or overlooked forest regions. These solutions to limited supply can therefore turn into problems by placing new demands on depleted forest ecosystems.

All markets are not equally available to all landowners. The opportunity to sell timber or other products from forestland depends on the availability of buyers. In some areas, the opportunity to sell low-grade trees for pulp or engineered wood products is plentiful, allowing for restoration-oriented logging to rebuild stand quality. In other areas there are no buyers and overstocked stands remain untreated. Conversely, increasing scarcity of large-dimension sawtimber has caused most mill owners to downsize their equipment to accommodate more typical logs. Having limited buyers and markets for products can depress prices and create disincentives for stewardship.

The Value of Forest Assets in the Marketplace

Markets still do not value forests per se; rather they value a relatively narrow stream of products, primarily wood, from forests. Alternatively, they value the development potential of forests. Most forest assets other than wood are poorly understood, poorly quantified, and typically overlooked within conventional economic valuations. Lack of alignment between the ecological and social values of forests and their financial or market value inevitably drives the simplification and loss of forest ecosystems as owners and investors seek to realize their financial goals. Given the force with which markets shape forests, the lack of financial incentives for conservation and stewardship of habitat, water quality, carbon sequestration, and other highly valuable forest assets is a great hurdle to overcome. Markets are beginning to emerge for some nontimber forest values, such as the ecosystem services of carbon sequestration and water provision. However, sophisticated market development and risk capital will be needed to develop markets for forest products that can rival the returns available from timber and development. Further, our scientific understanding of how to manage forest ecosystems for the provision of many products is still young. Expanding markets for nontimber forest products, recreation, or ecosystem services can unintentionally lead to the further degradation of forests. Finally, intrinsic forest values simply cannot be fully captured by the marketplace. If certain forest values are not financial, can they nonetheless be expressed by a social commitment to the conservation of forests?

Inequitable Flow of Forest Costs and Benefits among Stakeholders

Markets do not account for and allocate the costs and benefits of private forests equitably among all stakeholders. Private forests produce both private and public benefits. They also generate both private and public costs. The balance sheet of costs and benefits does not always add up appropriately. Some costs, such as the impact of sedimentation on water quality, flow to downstream landowners, to other industries such as fishing, and ultimately to the public. Some benefits, such as value-added profits or water supplies, flow to stakeholders that are very distant from the forest property and its community. These imbalances between cost and benefit create conflicts and claims that can sometimes be settled only through litigation, regulation, or political action. The feeling of inequity erodes social and political support for forestry as a desirable land use.

How the Web of Regulation Can Impede Stewardship

A complex web of federal, state, and local laws affect forest ownership and management. As public concern about the environmental impacts of private forest management on communities, neighbors, and public trust resources has grown, the number of regulations has grown, especially at the local level. This is another reflection of the increasing alienation between the public and the practice of forestry and between traditional timber-based forest owners and the growing number of amenity-oriented owners. When they are effective, regulations set the minimum thresholds for public acceptability of private activity, giving forest owners the "social license to operate." However, it is not always clear that the public benefit goals of species protection, improved water quality, and forest restoration are being achieved to the degree desired. Some laws have not been well enforced, others generate more paper than progress, and still others are so complex they invite noncompliance. Nonetheless, new laws related to forest practices continue to be added at the state and local levels as legislators respond to public demands. But do we have adequate incentives for forest owners to go beyond the minimum and restore biodiversity?

The activities of forest landowners fall under federal environmental legislation such as the ESA, Clean Water Act, and Clean Air Act; state forest practice, water quality, and endangered species laws; federal and state worker safety laws; state and local land use laws and zoning ordinances; as well as a patchwork of other local laws that vary from place to place. In all, as of 1992 twenty-one states regulated forest practices and forty-four states had instituted protection of endangered species in addition to federal protection. A study by the USDA Forest Service identified 644 nonfederal forest-related regulatory laws, 117 of which were at the state level and 527 at the local level (Greene and Siegel 1994). The North and South were found to have the greatest number of local laws, reflecting weaker state-level regulation, compared to the Pacific region, for instance. Most of the local laws have been passed in the last fifteen years. About 40% of all local ordinances regulate timber harvesting; 20% protect special features; and 20% protect the environment, addressing water quality, use of pesticides or herbicides, and other impacts of land disturbance. Other ordinances deal with protection of public property (roads, bridges), and protection of trees and shade. Municipal and county ordinances include those regulating forest practices as well as ones protecting greenbelts or proscribing hours for the operation of heavy equipment.

The challenge of compliance is made more difficult in that various agencies cover different resources and jurisdictions. At whatever level, they typically do not communicate well or coordinate permitting or enforce-

ment. At the national level, the NRC commented, "The federal administrative landscape of regulatory programs bears little relation to a holistic
approach to maintaining the integrity of forest ecosystems" (1998).

This checkerboard of regulation can be daunting for even the most
stewardship-minded forest owners. Further, the cumulative impact of
these laws and policies is not well understood and can have unintended
negative consequences for forest conservation and stewardship if, for
instance, needed habitat restoration activities cannot be undertaken without considerable time and expense invested in compliance with regulation.
In other instances, laws for achieving one environmental goal may impede
the accomplishment of another. For instance, requirements of the Clean
Air Act limit the ability of landowners to implement prescribed burning
programs on forestlands.

How Tax Policies Can Penalize Stewardship

Tax policies can have significant impacts on forests. Forests and forestry
are taxed in a variety of ways at the federal, state, and local levels. These
include federal and state capital gains and estate taxes as well as local
property taxes and timber yield taxes. Federal taxation in particular can
be quite complex, as forest management investments and revenue are
treated differently for different kinds of owners. Some of these taxes are
structured in ways that inhibit forest stewardship and conservation; or
they create incentives for forest conversion, fragmentation, or degradation. Some positive incentives—such as reforestation tax credits and special-use valuations for estate taxes—are used infrequently by NIPFs
owing to their complexity.

Capital gains taxes tend to penalize any long-term hold because they
tax not only real price gains but inflation as well. Therefore, the taxes fall
most heavily on forest landowners who invest in growing and managing
older forests over decades. For instance, a $1,000 asset becomes worth
$2,000 after twenty-four years of 3% inflation, although the asset itself has
not appreciated in any real way.

Another significant disincentive for NIPFs to undertake forest management is finding themselves classified by the IRS as passive investors
in their forests—as many do. This happens when NIPFs do not meet the
high statutory threshold of time spent managing their lands every year and
so cannot be considered active investors, because the demands of timber

harvest or other management activities are so episodic. Because of passive loss limits they cannot expense their annual costs and stewardship investments against other income as can larger forest landowners. Rather, they are forced to capitalize these expenses and cannot recapture that value until after a future (often distant) timber harvest.

According to the USDA Forest Service (1999), "Amortization of reforestation expenses, the reforestation tax credit, exclusion of cost-share payments to establish trees from gross income, deduction of management expenses, depreciation . . . and full casualty loss deductions are not available to the growing proportion of nonindustrial forest landowners who hold and manage their property primarily to provide environmental and social benefits." Limits on the ability of NIPFs to recover their forest maintenance and stewardship costs through tax treatments available to other landowners and enterprises seriously inhibit important investments in forest health and ecosystem restoration. Expenditures made by any private owner for forest stewardship activities that benefit the public interest, such as for habitat restoration, cannot be recaptured at all, whether through expensing or capitalization, because the IRS considers these expenditures to have been made without a profit motive.

Because of the appreciation of timber and land in recent years, forestland owners are much more likely than most other people to have sufficient assets to trigger estate tax liability (Greene et al. 1999). Currently assets of less than $675,000 (or $1,350,000 for married couples) are exempt from estate taxes. By 2006, up to $1 million per person will be exempt. Estate taxes for larger properties—and those with well-stewarded, mature forests—will be higher than those for cutover lands. If careful estate planning has not occurred and liquidity in the estate is inadequate, this can have the perverse effect of breaking up the property and/or forcing extensive harvesting of those forests to fund taxes. Estate tax rules and planning tools are complex and landowners with assets over the exemption limits must seek professional advice to effectively reduce their tax burdens. Most options for estate tax reduction require that action be taken before the death of the landowner, limiting postmortem actions by the heirs to reduce taxes and protect the land. (One important exception is the recently enacted option for executors to grant conservation easements on certain properties under IRC Section 2031(c).) The rules for special-use valuation that can limit the estate valuation of forestland are very complex and make it difficult for owners of working forests to meet all the requirements. With millions of acres of forestland going through the estate process over the next twenty years, unfunded estate taxes on larger, well-stocked forests could drive greater fragmentation and ecosystem degradation.

NIPFs frequently cite property taxes as a threat to their forest conservation efforts. Annual property taxes are a major component of the carrying cost of a forest and can be burdensome when they are high and the income of timber harvest or other revenue is infrequent. There are four basic property tax systems for forestland in the United States, according to an analysis by Sun Joseph Chang of Louisiana State University (1996). Regardless of the system, property taxes on the forest portion of a property usually range from $1 to $3 per acre, which is typically much lower than property taxes on other land uses. Seven states[1] have created special forest zoning, providing for low annual taxes based on a set forestland valuation (varying on site class) to promote forest conservation and management; an additional severance tax is due when timber is actually cut. Twenty-four states[2] have a tax system based on revenue-based forest productivity. Nine states[3] have a flat tax for all forestland regardless of productivity. Six states[4] exempt part or all of the forest value from property tax, although the underlying land can still be highly assessed, depending on its development value. Finally, seven states[5] still have an ad valorem tax system that taxes standing timber, creating a motivation to shorten rotations to minimize timber value on the ground. This latter system penalizes owners with older, mature timber. Many states require that landowners prepare forest management plans to qualify for reduced special-use property tax valuations. Typically, however, only a minority of

---

1. As of 1996 these included California, Georgia, Idaho, North Carolina, Oregon, Virginia, and Washington.

2. As of 1996 these included Alabama, Arkansas, Colorado, Connecticut, Florida, Georgia, Idaho, Illinois, Indiana, Kansas, Louisiana, Maine, Minnesota, Mississippi, Montana, Nebraska, New Jersey, Oregon, Pennsylvania, South Carolina, South Dakota, Texas, Utah, and West Virginia.

3. As of 1996 these included Kansas, Maryland, Massachusetts, Michigan, Missouri, North Dakota, Ohio, Vermont, and Wisconsin.

4. As of 1996 these included Alaska, Delaware, Indiana, Iowa, New York, and Ohio.

5. As of 1996 these included Arizona, Kentucky, Massachusetts, Minnesota, New Hampshire, Oklahoma, and Tennessee (Chang 1996).

otherwise eligible landowners have plans, which limits enrollment. Forest landowners are most worried about increasing tax burdens in states where forest property can be reassessed to reflect the increase in development values that accompanies creeping urbanization into forest areas. Reassessment can raise annual costs beyond the threshold that forest revenue can fund, leading to forest conversion.

Often property taxes raised on forest and farmland are used to subsidize services on developed land, as undeveloped land requires less public expense. The American Farmland Trust studied property taxes in the Northeast and found that for every dollar collected from undeveloped farmland and forestland, only $.33 was returned to them in services, whereas for commercial and residential land $1.15 was spent on services (1993). The following anecdote illustrates the dimishing municipal returns of forest conversion:

Dr. Green, a dentist, owned a 300-acre forest in the town of Hebronia for many years. There was no house on the land, and Green lived in another town. He visited and made occasional harvests but basically held the land because of sentiment (it was part of the farm where he grew up). The town collected $3,000 in annual property taxes from Green and spent $990 of this on local services relating to land uses such as forests, meaning that a "profit" of $2,010 from Green's taxes was available for services to the developed part of town. Unfortunately, Green died before making an estate plan that could have ensured his land passed on as a forest. The IRS estate appraisal found that, since there was road frontage on three sides of the woodlot, Green's heirs had inherited high-value development land. The heirs, lacking cash to pay the assessed values, sold the land to a developer who subdivided the forest into large estate lots and built thirty fine houses on the land. The town now collects $120,000 per year in taxes from the thirty new households but pays $138,000 per year for services demanded by the development. This is an $18,000 "loss" in net tax income compared with the $2,010 "profit" in net tax income from the formerly undeveloped forest. The town is collecting forty times more in taxes . . . and losing money on the deal.

Why We Know So Little about Our Private Forests

Despite the prominence of private forests on our national landscape, we know very little about their extent and condition. Without better data and analysis, we simply cannot see the forest for the trees. While data on habitats, water quality, soils, carbon stores, forest health, and ecological function are poor or nonexistent, timber data are often not of high quality either. Changing forest conditions are not well documented. Government data sets are often out of date and incomplete and lack integration across regions, both within departments and across departments and jurisdictions. Data that do exist are often inaccessible or incomprehensible to the layperson. How can we foster greater forest conservation and stewardship if we do not know what the forest conditions and trends are, especially at the regional and subregional levels?

Equally important, forest landowners are not well understood. Our knowledge of who owns forests and why is relatively rudimentary. As with forests, so much of the available data deal with a narrow set of issues, all revolving around timber stocking and harvest. Other questions regarding the attitudes, needs, and interests of landowners in relationship to their forests have been lacking. There are a lot of myths about forest owners— especially about the millions of individual owners. Broad characterizations are often based on small samplings that include larger and more activist owners, missing out on the broader spectrum of ownerships. How can we better engage forest landowners in conservation and stewardship when we do not know more about who they are, what motivates them, and how to reach them?

Forest management and silviculture that seek to restore and enhance forest ecosystem functions are still in early stages of development. We do not adequately understand many natural forest types and situations. Further, we have not translated the ecological forest management knowledge we have into user-friendly, multiresource, decision-support systems to help managers weigh the impacts of their options. Without better applied science in forests, forest degradation and loss will continue apace.

Increases in Natural Threats and Frequency of Disturbance

As described in chapter 2, natural threats to forests are on the rise, fueled to new levels by human influences. Fire suppression has perversely increased fuel loads and the threat of hotter, larger, and more catastrophic fires. Pests are producing more extensive damage in stressed and simplified forests. Roads and other development are vectors for the introduction of invasive organisms that damage ecosystem functions and threaten native species. Acid rain and increased ozone in the lower atmosphere are damaging forest metabolic functions. Climate change brings with it huge unknowns about forest response to increased CO_2 levels, changing growing conditions, and the increased incidence of intense storms. Further, anthropogenic disturbances are more compressed in time compared with past cycles of natural disturbance, stressing forests more and giving them less time to recover structure and functions. Any one of these natural threats poses daunting risks for U.S. forests.

Who Cares about Forests?

Unfortunately, this litany of threats to private U.S. forests is not of great concern to most Americans. Yet without better engagement of public interest in their fate, private forests will continue to diminish in extent and quality. The growing lack of connection between people and forests is perhaps the greatest barrier to increased public and private interest and investment in forest conservation. Our primarily urban population has little contact with forests outside of nearby parks or in wilderness settings. Those few who do spend time in rural forest areas tend to be higher-income urbanites without ties to traditional forest uses. In fact, they are often drawn to forest areas as retreats, investing in recreational or second-home properties and thereby adding to fragmentation. Owners of smaller, forested parcels view their trees as amenities, not major elements of an ecosystem.[6] Meanwhile, rural families are having difficulty involving their own next generations—often urbanized professionals themselves—in the management of family forests. Lack of interest among offspring was the top barrier to conservation of family forests cited by participants in the NIPF survey conducted for this book by Mater Engineering.

Forest products comprise an ever smaller portion of the U.S. economy, yet they appear plentiful and inexpensive to consumers. The dynamics and impacts of production and consumption are remote from most Americans. The complexities of forests and forest management, daunting even

6. One indicator of the amenity value of forest fragments in residential settings is the amount people spend on tree care, rising 50% to $2.4 billion from 1996 to 1997, with forested homesites representing 27% of the market. This is more than twelve times last year's annual USDA Forest Service budget for the Forest Stewardship, Forest Improvement Program, and the Inventory and Analysis Program (Sampson 1999).

to landowners and resource managers, are not grasped by the public. In general, the public does not like logging. Forestry is not a trusted profession. The forest issues that have held the attention of national media have been oversimplified, creating such polarities as "owls versus jobs" and other clichés. Little real dialogue occurs even among those most concerned about the future of America's private forests.

Lack of public interest and support for investment in the stewardship of private forests is reflected in the allotment of less than 1% of federal funds for all private forest-related programs. Forest landowners, while potentially representing a major political constituency supporting forest conservation, are so varied and dispersed they are difficult to organize and unite. Established landowner and forestry associations have not succeeded in creating broad-based memberships of owners, professionals, and stakeholders. No one has been able to tell the story of the crucial services private forests provide to our urban, diversified, postindustrial society, nor the story of how important landowners are to the conservation of private forests for the public benefit.

Part Two

The Conservation Toolbox and How to Use It

TO ACCELERATE THE CONSERVATION of U.S. private forests, the array of threats and barriers discussed above must be prioritized, addressed, and resolved as well as possible. Therefore within a coherent strategy an assortment of tools are needed, appropriate to different aspects of the problem. In the following chapters we provide an overview of the major tools in our toolbox, referencing specific projects and initiatives wherever possible. Although we have not engaged in an exhaustive discussion of each and every conservation initiative, we believe we have broadly captured a rich range of proven and emerging approaches. A contact list for organizations mentioned is included in appendix C. In the final chapter we will prioritize these tools to best address the key threats and barriers, forming a strategic road map to more effectively conserve private forests. To begin, we will look through the different compartments of the toolbox:

Public policy tools (chapter 4)
  Federal and state regulations and incentives
  Political organizing
Cultural tools (chapter 5)
  Scientific research and information
  Communication and education
  Technical assistance
Market-based tools (chapter 6)
  Fiscal incentives
  Financial mechanisms
  Market development

Each area contains existing, successful applications of these tools—such as government cost-share programs and conservation easements—as well as new tools just being identified and developed to address the challenge of maintaining and restoring America's private forests. These include new forms of ownership, new institutions for stewardship, new business models of forestry, new financial instruments and funding sources, and new conservation partnerships among private, public, and nonprofit entities. Different tools will have applicability for different audiences and issues, and will engage different providers or partners from different sectors.

As we discuss in the final chapter, to fully utilize these tools to stem the loss of America's private forests, greater understanding and investment will be required by a range of stakeholders: forest owners of all kinds, non-governmental organizations, public agencies at all levels, natural resource professionals, elected officials, and the general public. Our society will need to complete the transition from valuing forests just for timber or for land or even for parks. Fundamentally, to protect America's private forests over the long term all concerned need to work together to build a culture that values forests as forests, regardless of ownership.

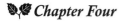 *Chapter Four*

Public Programs and Policies for Private Forests

Public policy can foster private forest conservation, just as it can be an impediment. Building public investment in the conservation of private forests is critical to preventing continued loss of valuable forest resources that benefit society. More effective implementation of current programs as well as creation of new incentives can motivate more landowners to make long-term stewardship investments. In this section we will provide an overview of the history and status of state and federal programs relating to private forests. We will also point out public policy developments and evolving tools related to improving the effectiveness of regulations, expanding fiscal incentives, and organizing more public support. It is especially important for the general public to gain a greater *sense of ownership* of these forests if they are to make more investments in them along with private forest owners.

Federal Programs

Although federal investments in private forests are dwarfed by many state programs, federal programs influence the direction of states and are important incentives for stewardship. In general, federal programs are implemented cooperatively with state agencies. In the initial stages, the federal government's involvement in forestry programs aimed at private forests involved two cabinet departments—Agriculture and the Interior. Until very recently forestry was generally perceived as meaning timber production, and therefore as an agricultural activity; consequently, the programs and agencies rapidly coalesced within the U.S. Department of Agriculture (USDA), where they remain to this day. The major USDA forestry agency is the Forest Service, but other agencies have had important roles over time. Some, such as the Natural Resources Conservation Service (NRCS), are positioned to greatly assist programs in the future. Wildlife-oriented landowner assistance programs, however, still reside in the U.S. Department of the Interior under the U.S. Fish and Wildlife Service (USFWS). Endangered Species Act (ESA) compliance is split between the USFWS and the National Marine Fisheries Service, part of the U.S. Department of Commerce's National Oceanic and Atmospheric Administration (NOAA). Over the last thirty years in particular, the role of the U.S. Department of the Interior has become relatively more prominent, as the U.S. Fish and Wildlife Service has provided increasing assistance to forest landowners for habitat stewardship. The National Marine Fisheries Service recently emerged on the private forest regulatory scene with the listing of salmon species under the ESA.

The evolution of USDA programs to encourage forest stewardship on private lands has created a complex mix of approaches, all of which involve more than one agency and more than one level of government. The primary federal agencies involved in forestry programs include the USDA Forest Service and the USDA Cooperative State Research, Education and Extension Service (CSREES). Aside from general regulation (tax and environmental laws), the focus of federal forestry programs includes providing information (for example, education, workshops, publications, research) and voluntary services and incentives (technical assistance, cost-sharing, loans; see table 4-1).

Educational programs administered by the U.S. Department of Agriculture include land grant universities and state extension services; technical assistance programs involve state forestry agencies (SFAs); and cost-sharing programs may involve both the local offices of the NRCS and the Farm Services Agency (FSA), as well as the cooperating local Resource Conservation Districts (RCDs) and the locally elected committees that help administer FSA programs.

In 1862, although forestry was not yet a national concern in the United States, Congress established the Land Grant College System through the Morrill Act. This provided each state with a grant of 30,000 acres from the public domain (or its cash equivalent) for establishment of a college of science. Thus the practical arts such as agriculture and forestry (the art of growing and harvesting trees) would be taught and studied at the same academic level as the rest of the programs in America's growing university system.

In 1880 Congress passed a bill to make forestry a permanent part of the USDA, with a Division of Forestry assigned to focus primarily on developing and publishing information about forests and the forest industry. When Gifford Pinchot was named to head the USDA's Division of Forestry in 1898, he was dedicated to the idea of changing it from a bureau of information that produced only pamphlets and reports into an active, on-the-land participant in practical forest management. The main target of concern was private lands, almost entirely in the hands of farm families. The challenge that drove Pinchot—and the emergent USDA Forest Service—was to help those landowners learn and implement better timber management in private forests.

In the 1930s President Roosevelt called for quick authorization of a Civilian Conservation Corps (CCC) that put people to work immediately planting trees, stopping gullies, fighting forest fires, and building dams. In the nine years of CCC operation about 3 million men were involved, for a total cost of some $2.5 billion (Zimmerman 1976). Some of their impressive forestry accomplishments included construction of more than 1,300 fire lookout towers and more than 50,000 miles of roads and trails, planting of 1.25 billion trees, and more than 2 million man-days of fire fighting. The 1930s also saw the creation of a number of federally administered conservation programs and services including the Prairie States Forestry Project, the Land Utilization Program, the Soil Conservation Service, and the Norris-Doxey Farm Forestry Act.

After the end of World War II, the partnership between the USDA and the state forestry agencies expanded its capacity to provide technical assistance to landowners. The Cooperative Forestry Act in 1950 strengthened funding support, and by 1960 forty-six states and Puerto Rico had more than five hundred field foresters working with woodland owners.

The decade of the 1970s saw a great outpouring of national legislation that affected forestry and conservation including the ESA, the Clean Water Act, and the Clean Air Act. The nation was grappling with the realization that its postwar growth and technologies were creating enormous problems that threatened to ruin the land, water, and living resources on which society rests. In a steady drumbeat of bad environmental news,

Table 4-1.
USDA programs and services available to NIPF landowners

Program	Function	Administering Agency	Field Delivery Agency
Environmental Quality Incentives Program (EQIP)	Cost-sharing	NRCS, FSA	FSA, NRCS, SFAs
Conservation planning	Technical assistance	NRCS	NRCS, RCDs
Conservation Reserve Program (CRP)	Cost-sharing, land rent	FSA, NRCS	FSA, NRCS, SFAs
Forest Legacy Program	Easements to protect forestland uses	FS	FS, SFAs
Forestry Incentives Program (FIP)	Cost-sharing	NRCS, FS	SFAs, NRCS
Forest Stewardship Program (FSP)	Planning and management technical assistance	FS	SFAs
Limited resource farm loans	Loans to low-income landowners	FSA	FSA
Resource Conservation and Development (RC&D)	Technical assistance, grants, loans for community projects	NRCS	NRCS, FS
Small Watershed Program (PL-566)	Cost-sharing for watershed projects	NRCS	NRCS, FS
Stewardship Incentives Program (SIP)	Cost-sharing	FS	SFAs, NRCS, FSA
Renewable Resources Extension Act (RREA) and Smith-Lever funding	Education	CSREES	extension services

Source: Sampson and DeCoster 1997

people became aware that poisoned rivers, toxic air, and degraded land-scapes needed immediate attention, and that letting individual communities and states address the issues in their own ways and at their own pace was not going to meet the challenge. A new generation of forest practice acts was passed in many states, including comprehensive acts in the major timber-producing states of California, Oregon, Washington, and Idaho.

The passage of the National Environmental Policy Act (NEPA) in 1969 focused attention on the need to consider the broad environmental impact of major public actions. The Federal Water Pollution Control Act was amended in 1972 with a major new focus on the prevention of pollution from soil erosion and sedimentation. An approach to nonpoint-source pollution based on best management practices (BMPs) was proposed that could be designed by natural resource specialists and adapted to local conditions.

In the process, government education and technical assistance programs for NIPF landowners came to incorporate pollution prevention (for example, sedimentation) as an integral part of forest management. Extension foresters devised educational programs to reach NIPF landowners with information about their responsibilities for pollution control, and the best methods for achieving it.

While targeted at stabilizing highly erodible cropland, the Conservation Reserve Program created under the 1985 Farm Bill has had a major impact on forestland by promoting the planting of more than 2.5 million acres of trees—more than either the CCC or the Soil Bank of the 1950s (Sampson and DeCoster 1997). Ninety percent of the reforested acres were in the South, probably restoring previously cleared forestland.

The 1990 Farm Bill contained the first forestry title in the history of the Farm Bill, and departed from the traditional short-term nature of farm programs by creating permanent forestry programs. This became the basis for most of the current forestry programs important to NIPF landowners, including the Forest Stewardship Program (FSP), Stewardship Incentives Program (SIP), and Forest Legacy Program.

At the core of the USDA private forest programs are several that assist in the management and conservation of forestlands when the landowner voluntarily cooperates with the government. These are targeted primarily toward NIPF owners, often by restricting eligibility to them or setting maximum acreage or dollar limits. Over the last decade, more federal assistance has been targeted at landowners' multiresource forest stewardship and conservation activities than in the past, when tree planting was often the major focus. These landowner programs include educational programs, technical assistance, and financial incentives.

USDA educational programs help landowners understand forestry principles, public policy issues, and/or forestry program options. Education means conferences, workshops, tours, classes, and the production and distribution of educational materials that are directed at groups of people rather than toward individuals (Baughman 1993). State forestry agencies carry most of the responsibility for implementing the federal programs administered by the USDA Forest Service through the Cooperative Forestry Program. Universities and the Cooperative Extension Service play a large educational role often funded, at least in part, with pass-through financial support. RREA funding through CSREES may leverage several funding sources together into one educational program. NRCS is also involved in educating landowners, in cooperation with local soil and water conservation districts. For several of these programs the distinction between education and technical assistance is somewhat blurred, as the two activities are often conducted simultaneously.

Technical assistance—usually on site—helps landowners with the development of management plans for their land, as well as in solving specific problems. (Technical assistance is also provided by the private sector, through consulting foresters or as part of landowner assistance programs operated by some industrial companies.) The Forest Stewardship Program (FSP), a major section of the 1990 Farm Bill, encouraged landowners to develop Forest Stewardship Plans that protect all the resource values of their forestlands. These plans, prepared in most states with the assistance of a state or consulting forester, provide the basis for qualification in cost-sharing under the companion Stewardship Incentives Program (SIP).

Financial incentives help offset the costs of applying forest management or conservation measures. These are usually cost-sharing, low-cost loans, land rent, funding of conservation easements, or other public assistance aimed at specific practices such as tree planting. In combination, USDA forestry cost-share programs have assisted almost 40% of all the NIPF tree planting in recent years (41% in 1995, for example), so the significance of these incentives cannot be minimized (Moulton et al. 1995). There is also good evidence that in the southern region, where NIPF owners hold the majority of the productive pine forestlands, both cost-sharing and tax credits are important in achieving reforestation after harvest, particularly among marginal managers (Royer and Moulton 1987).

The Forest Stewardship Program is a program that integrates planning, technical assistance, cost-sharing for activities, and establishment of conservation easements. As this program emphasizes overall stewardship over a simpler focus on tree planting or timber harvest, it is especially appealing to NIPF owners with their multiple ownership goals. As a package, these FSP elements provide a suite of incentives for interested private landowners. Once a landowner has completed an FSP, then an SIP (if it or a similar program is reauthorized and funded by Congress) can offer cost-sharing on such practices as wildlife habitat enhancement and riparian area protection in addition to the traditional assistance in tree planting and timber stand improvement. Landowners with FSPs are also eligible for participation in Forest Legacy, through which a permanent conservation easement is sold or donated to the USDA Forest Service, state agency, or, in some instances, nonprofit land trusts. The purposes of Forest Legacy easements are to prevent fragmentation and conversion of productive forestland, maintain traditional land uses, and protect significant environmental values on private lands. The easements must be purchased from willing sellers, at fair market value. The federal share of program costs cannot exceed 75%, with the remainder coming from state, local, and private sources. States enroll individually in Forest Legacy based on a state-produced Assessment of Need. The program's funding has been modest, even with an increase to $30 million in the 2000 budget. To date, twenty-two states and territories are participating and nine others applying to join. As of April 30, 2000, the program had protected more than 111,000 acres in eleven states, with a value of $54.6 million (USDA Forest Service, 2000). The program is increasingly popular with landowners and has a backlog of projects proposed for another 330,000 acres of private forestland.

Another integrated initiative created in the 1996 Farm Bill was the Environmental Quality Incentives Program (EQIP), administered by NRCS at the national level. EQIP provides farmers and ranchers—including NIPFs—with educational, financial, and technical assistance to address threats to soil, water, and other natural resources. Local districts establish priorities for conservation and restoration funding. EQIP encourages a variety of forestry-related practices such as riparian restoration, erosion control, water quality improvements, and wildlife habitat management. When EQIP was created, two older tree-planting-oriented efforts, the Agricultural Conservation Program and the Forest Incentives Program, were folded into it. The Farm Bill also created the Wildlife Habitat Incentives Program, again administered by NRCS, to provide technical assistance and cost-sharing payments to landowners to establish and improve fish and wildlife habitat.

The USFWS also provides technical and financial assistance to forest landowners through the Partners for Fish and Wildlife Program. Similar to USDA programs, this is a voluntary cooperative effort to restore important habitats for federal trust species, for example, migratory birds, threatened and endangered species, anadromous fish, and certain marine mammals. This program gives priority to projects on permanently conserved land, and to those that reduce habitat fragmentation; conserve or restore threatened natural communities; result in naturally sustaining systems; and/or are identified by local and state partners as being of high priority. Since 1987 this program has worked with more than 19,000 landowners to restore 40,174 acres of wetlands, 333,165 acres of native prairie and grassland and 2,030 miles of riparian and instream aquatic habitat.

In addition to the landowner service programs, the USDA Forest Service engages in research, inventory, forest health monitoring, and natural hazard protection activities that benefit all forestlands, public and private, in large and small ownerships, whether the landowner cooperates or not. They include:

- Information, monitoring, and data reporting programs that provide an ongoing assessment of forest conditions and trends in the nation, market conditions and trends, and other environmental facts and conditions. The prime example here is the Forest Inventory and Analysis (FIA), whose data are relied on by virtually everyone in both forestry and the forest industry in America. To provide more timely data on forest conditions, the FIA is moving to a program of continuous inventory monitoring, and intends to improve its Web-based data access accordingly.
- Research that develops scientific understanding of forest systems and how they function; improved methods of management; new technologies, equipment, and processes; and new products that improve use of forest resources.

Figure 4-1.
Federal funding for USDA forestry technical assistance programs

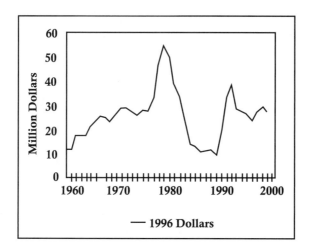

Figure 4-2.
Federal funding for USDA forestry financial assistance programs

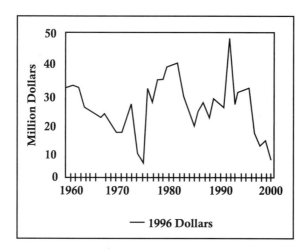

- Seedlings, nursery, and tree improvement programs that help states upgrade the quality of their nursery operations and produce trees that are locally adapted.
- Programs where federal, state, local, and private cooperators provide fire protection and enter cooperative efforts to reduce damages from insects, diseases, and other threats to forest productivity.

The State of Federal Funding for Landowner Assistance Programs

In 1997 the State and Private Forestry unit of the USDA Forest Service, which is responsible for most of the USDA Forest Service landowner programs we have described, had a total budget of $136.8 million, or 6% of the total USDA Forest Service budget (NRC 1998). Although in general the USDA programs have proven popular and effective with forest landowners, federal appropriations have fluctuated considerably over the last forty years, as illustrated in figures 4-1 and 4-2. While the number of smaller NIPFs who need technical assistance has more than doubled since the 1970s, current funding has not risen at all. For landowners seeking financial assistance, a key stewardship barrier for smaller NIPFs, funding has plummeted to less than a quarter of what it was ten years ago.

State Programs

Forest practice regulations and stewardship programs are implemented at the state level, even if they originate as federal programs. Different forest types and conditions, political systems, land ownership patterns, timber market opportunities, and local customs all contribute to the diversity of today's state forestry programs. Therefore, the United States is a mosaic of state programs, influenced by federal policies and programs, but significantly different from state to state. State-level programs have the potential to be more relevant to the issues of local private forests and their owners. Regardless of their differences, all states utilize the same basic set of tools to promote forest stewardship: regulation, education, technical assistance, and financial/fiscal incentives (with most emphasizing technical assistance and education). In general, regulation is used to set minimum standards for forest practices to ensure protection of public trust resources on private lands, while incentives are used to encourage and enhance stewardship above and beyond the thresholds established in regulation.

Cooperative, incentive-oriented programs seek to protect private forests from wildlife, insects, and disease, and to improve the quality and abundance of timber, water, wildlife, recreation, and other forest values. See appendix B, table B-10, for a regional overview of the tools used by the states to accomplish their major forest resource objectives.

Historically, state programs, in cooperation with the federal government, focused the bulk of their resources on preventing wildfires and encouraging reforestation. State forest practice regulations have evolved considerably from the first laws passed during a period of intense national debate over logging impacts in the 1920s. Rather than face federal regulation and increased federal forest ownership, states began to take steps to improve forest management. By 1950 some form of forestry regulation covered about 20% of U.S. timberlands. With the explosion of public environmental concern in the 1960s and 1970s, a new generation of state laws began to be passed, as mentioned above. As of 1992, ten states, mostly in the West and New England, had comprehensive forest practice acts, seeking to unify the patchwork of local regulations while addressing a broader range of environmental impacts than earlier rules oriented toward timber stocking. Another twelve states are moving in the direction of similar comprehensive acts. If this occurs, nearly 40% of U.S. timberland will be regulated in this centralized fashion—more than all public timberland together. Other states disperse aspects of forestry regulation among a variety of agencies. Southern states have resisted the use of regulation, preferring voluntary or quasi-regulatory BMP programs combined with cost-sharing assistance. With the increasing intensity of forest management, especially in the South, more states are considering moving toward a regulatory model (Ellefson et al. 1995).

In general, the forces driving increased state forest regulation include public response to poor forest practices; heightened public interest in environmental protection; compliance with federal environmental law; public desire for accountability; proliferation of local ordinances; landscape-level forestry concerns; and the desire to emulate the actions of other states. Opposition to the regulatory approach is based on the resistance of landowners to increased regulatory compliance in general; increased costs to landowners; the added complexity and cumbersome nature of regulation; as well as the high public cost of administering such programs (Ellefson et al. 1995).

Although there is much debate over the effectiveness of regulation versus incentives, the NRC notes: "In general, however, the effect of public regulatory programs on non-federal forests has been to increase tree planting, improve water quality, and protect wildlife habitats. Whether these results would have been greater under a different type of program (for example cost-share or technical assistance) is not known. Analyses showed that in Oregon, Washington and California, for instance, 30 to 40 percent, 10 percent and 25 percent more area, respectively, was reforested since the inception of each state's forest practice regulatory law. . . . Other assessments have shown that current and expected state regulatory programs would often substantially increase private timber inventories over base (or expected) inventories, with harvest volumes remaining the same or slightly different from base levels" (1998).

The strategic leadership in NIPF forestry programs has largely shifted to one where the states work more closely with the USDA Forest Service. One of the major factors in the shift was the widespread launching of state forestry plans in 1978, most of them completed during the 1980s (Gray and Ellefson 1987). That effort continues to provide the information and impetus for an ongoing strategic planning process in most states.

In 1994 state agencies invested $1.1 billion in forest-related programs, an increase of almost $500 million from 1984 levels (NRC 1998). Although state forestry programs have always struggled for adequate funding, even with declining federal funds, few state programs today depend on federal sources for more than 15 to 25% of their NIPF forestry activities. Federal funding is important, however, in introducing new concepts such as forest stewardship and in influencing trends across state and regional boundaries. Many states are turning to innovative funding sources such as market-based programs, fees, and voluntary contributions (Hacker and Baughman 1995). The more important these sources become to an agency's programs, the more entrepreneurial the agency must become, creating partnerships among other agencies, nonprofit private organizations, and private businesses to accomplish its goals.

Some of the most important differences among state programs can be traced to the personal traits of the state's program leaders over the years. In New Hampshire, for example, the state forester and the extension director entered into a working agreement in 1925 that continues to this day, reviewed and amended regularly to recognize new programs and needs. Setting the stage for close cooperation from the early days no doubt started the program on a different trajectory than if those two individuals had seen themselves as adversaries.

Today a number of state and local programs exist in which interagency relationships reflect the personal and professional relationships among the agency managers as much as or more than agency policies. That dynamic can change with one personnel change—either for better or for worse. The public approach taken in state programs affects the private sector as well. The forestry program leader in New Hampshire feels that the state's program of providing educational assistance, rather than providing services to landowners, has encouraged the development of a thriving private-sector consulting business (Edmonds, pers. comm. 1997). New Hampshire forests are commonly smaller parcels and are often owned by people with goals other than timber—a situation normally thought not to favor the development of a strong private-consultant sector. Yet this small state has more than 120 licensed professionals successfully earning their livings as forest consultants.

At the state level, public programs for forest conservation and stewardship can also benefit from more readily building broader coalitions and improving people's ability to understand conditions within the state. This has almost certainly contributed to the development of regional forestry coalitions such as the Lake States Forestry Alliance (Michigan, Minnesota, and Wisconsin) and the Northeastern Forest Alliance (Maine, New Hampshire, Vermont, and New York), where common interests can be addressed.

To remain relevant and improve effectiveness, future federal programs will need more than ever before to focus on state or regional-level strategic guidance and build in more flexibility for state-level implementation. From the federal point of view, the existence of increasingly strong state strategic capability helps argue the point that state-directed programs can be depended on to effectively achieve national goals and objectives.

The Effectiveness of Public Policies and Programs

Numerous studies of program managers and landowners have been conducted to assess their perception of the relative effectiveness of these approaches in accomplishing their stated objectives. Although results are mixed, depending on the location and characteristics of the respondents, in general it appears that all of the tools are considered effective and that educational, technical assistance, and tax policies generate the most support from landowners and agencies, while regulation generates the least. Forest regulations are not popular with most forest owners; nonetheless, they have been reasonably effective in increasing reforestation, improving water quality, better protecting threatened and endangered species, and raising standards of forest practices in many states when voluntary approaches proved inadequate. Clearly, a mix of public-sector tools is necessary to encourage greater landowner conservation of forests while ensuring a bottom-line level of protection acceptable to society.

Some studies have found that financial returns to owners were affected more favorably by cost-sharing programs than by tax programs, even if owners took advantage of all available tax benefits (Campbell 1988). However, a broad array of landowner surveys indicate that tax policy is very important in guiding decisions (McColly 1996). Sampson and DeCoster's "mini-survey of program insiders" indicated that "professionals deeply involved in forestry programs" felt that tax policies that reward landowners for investing in desirable practices motivate landowners the most (1996). They also voiced strong support for education and technical assistance. On the other hand, a survey by Ellefson et al. (1995) of program managers rated the effectiveness of education and technical assistance over voluntary guidelines, tax incentives, and regulation.

Education and technical assistance usually focus on encouraging landowners to better understand their forests and develop forest management plans. One study of woodland owners in Minnesota found that when owners had management plans, they were more likely to carry out other forest management activities (Rathke and Baughman 1996). Because management plans are almost always the result of technical assistance, Baughman (1993) concluded that technical assistance appeared to be the most effective incentive. Support for that contention was also derived from a North Carolina study that concluded that technical assistance had more impact than cost-sharing on the amount of tree planting that was accomplished (Boyd 1983).

Since tree planting has been the focus of most federal incentives, the results of a study by Alig et al. (1990) on landowners' tree planting are worthy of highlighting: (1) cost-sharing resulted in more tree planting; (2) cost-share plantings were not usually liquidated at the end of the cost-share contract; (3) technical assistance during harvest tended to increase stumpage revenue for owners and resulted in improved residual timber stands; and (4) technical assistance was correlated with more timber harvesting (NRC 1998).

In interviews with Advisory Group members and others for this book, the authors found strong support for the Forest Stewardship Program in particular. Interviewees usually also decried the elimination of funding for the Stewardship Incentives Program.

Improving the Effectiveness of Forest Regulations

Forest regulation has gone through several phases in this country, and it may be entering a new one as policymakers and the forest community grapple with how to improve compliance and then go beyond the minimum levels established by law to create incentives for expanded conservation and stewardship. The tension between prescriptive and performance-based approaches to forest regulation is high—just as it is in, say, forest certification systems. The new trend in forest regulation may be to combine higher thresholds of regulatory protection with new incentives for meeting and exceeding those thresholds. Incentives within the regulatory context usually provide the landowner with some combination of (1) assurances of a period of regulatory stability (such as through long-term permitting possible with HCPs under the ESA or through California's Non-Industrial Timber Management Plan available to forest owners of less than 2,500 acres); and (2) flexibility in meeting higher levels of performance goals, as with the new Safe Harbor Agreements under the ESA. Other incentives that seek to achieve the same goals as regulation include many of the cultural and financial tools discussed in this section, as well as the fiscal tools identified above. Overall, a greater investment is needed on the part of the forest community to examine the successes and failures of forest regulation; identify unintended negative consequences for forest conservation and stewardship; and devise new tools for engaging the forest owner's self-interest in protecting public trust resources on private lands. Three examples of current initiatives are outlined below.

State Incentive Committees
As the states in the Pacific Northwest increase regulatory restrictions on a host of land use practices based on the ESA's listings of several salmon species, some others have initiated state-level stewardship committees to advise their forestry departments and policymakers on taking a dual-pronged approach to the challenge. For instance, the state of California has been seeking to improve compliance of its forest practice

regulations with the Clean Water Act's nonpoint-source pollution requirements and with the ESA's requirements against the taking of threatened
salmon. At the same time as the California Board of Forestry considered
new sets of rule packaging, the state Resources secretary convened a
Forestland Incentives Task Force that includes a range of forest stakeholders to review and recommend a set of incentives to encourage landowner investment in water quality and salmonid protection. The administration can then take recommendations from the task force to the
legislature with greater stakeholder support than might otherwise be
anticipated. Although useful, the effectiveness of special task forces is limited if the committee is disbanded after its report is issued, limiting follow-
through on implementation. Institutionalizing stakeholder-based advisory
groups may be more successful, as described next.

Multistakeholder Forestry Councils
Established under the Sustainable Forest Resources Act of 1995, the Minnesota Forest Resources Council is responsible for developing and implementing the programs created by the act; members also serve as advisors
to state and local government on sustainable forestry. The council com-
prises fourteen members representing the breadth of forest stakeholders
in the state. Unlike forest practice regulatory boards in, for example, California or Washington, the Forest Resources Council combines the suite
of public programs under one citizen body, potentially allowing for greater
coordination and a more holistic approach to the challenge of forest sustainability. The council has been responsible for creating an integrated program to achieve sustainability of the state's forests. It includes (1) creating
comprehensive, user-friendly guidelines for forest management practices
that minimize negative environmental impacts; (2) implementing landscape-level planning across ownerships in various regions of the state; (3)
establishing a suite of forest resource and forest practice monitoring programs; (4) overseeing the Interagency Information Co-operative to coordinate the development and use of forest resource information in the state,
including developing common database formats and Web site access; (5)
creating a Research Advisory Committee to identify priority forest research, foster communication among researchers and users, and provide
funding; and (6) providing continuing education opportunities for professionals and the public, including the Logger Education Program, which
trained nearly eight hundred loggers in 1998 and the Institute for Sustain-

able Natural Resources, which brings current research, new technologies, and state-of-the-art practices to natural resource managers. *The Biennial Report* on the implementation of the SFRA is a readable document available to the public that clearly reports on the extent and condition of Minnesota's forests and relates vision and goals the Forest Resources Council has established for its work (MFRC 1999).

Local Land Use Planning

Land use regulation happens at the local level to a greater or lesser degree, although some state governments such as Oregon establish an overarching framework for local government decision making. In every forest region a new generation of land use planners is seeking to come to grips with the impact of sprawling nonforest uses and fragmentation. Much more support needs to be given to the many rural forest counties in sprawl's "expansion market." Planners, elected officials, and the public need to be educated about their options in protecting forestland, forest resources, and forest uses. They need to be able to fully consider the use of urban growth boundaries, exclusive forest use zoning, cluster zoning, transfers of development rights, conservation easements, infrastructure development fees, and other means to channel growth while minimizing conversion and further fragmentation. This dovetails with local efforts to finance forest conservation through real estate transfer taxes, sales tax increases, and other means.

The Sierra Business Council (SBC), based in Truckee, California, is an association of five hundred business members working to secure the economic and environmental health of the Sierras. SBC is working to help communities and planners understand the shift in the Sierra's resource-based economy wherein environmental quality is key to future economic growth. SBC's Sierra Leadership Seminar is a county leadership program for the Sierran counties—many of which are among the fastest growing in the state. SBC also sponsors the Planning for Prosperity seminar for local planners and decision makers with the Regional Council of Rural Counties, the Local Government Commission, and the local chapters of the American Planning Association. These programs help key people build their skills and understanding of appropriate tools to tackle the complex problems of planning in resource-based communities. Working with Placer County, SBC brought together cutting-edge ideas from around the country to help develop a start-of-the-art open space, farmland, and habitat protection program as part of its updated County General Plan. More such initiatives are needed not only in California but around the country.

Improving Landowner Compliance with the ESA

As discussed elsewhere in this book, the number and acreage of HCPs produced by landowners to guide their compliance with the ESA and provide them with a permit for incidental take of listed species are exploding. Most of the larger plans are for forestlands. Although this trend is positive for landowner engagement in the protection of threatened and endangered species, it also poses issues for adequate public review of these typically twenty-year-plus permits. Much of the science of forest ecosystem management is still young, so HCPs may represent the largest on-the-ground implementation of certain precepts to date. This is the basis, then, for scientific disagreement over the terms of HCPs, as was evident in the groundbreaking case filed by the Pacific Lumber Company for its redwood forests. For both scientific and procedural reasons, many environmental groups have opposed the current HCP process. Conservation and environmental advocacy organizations across the country provide important services such as timely independent scientific review and public comment on pending complex HCPs.

Safe Harbor Agreements under the ESA are an innovation of the Clinton administration to remove certain disincentives for private owners to manage their lands for the benefit of listed species. Available on a limited basis so far, the Safe Harbor Agreement aims to alleviate landowners' fears that if they successfully maintain or enhance habitat for listed species on their property, their activities may subsequently be restricted by the act. Using a Safe Harbor Agreement with the USFWS or NMFS, a landowner can manage to create, restore, or maintain habitat or otherwise benefit listed species without incurring ESA restrictions as long as the property maintains a certain baseline or target condition and species population. The site-specific agreement is formalized by the issuance of an enhancement of survival permit that authorizes take above the baseline or target condition. These agreements are appealing because they provide landowners with greater management flexibility and regulatory certainty for their activities.

Safe Harbor Agreements, such as the one concluded in 1999 by the Norfolk Southern Corporation on a 15,000-acre forest in South Carolina, may be especially useful in the protection and expansion of habitat for the endangered red-cockaded woodpecker because active management, such as prescribed burning and other control of understory vegetation, is important to maintaining the species' preferred forest structure. Without a Safe Harbor Agreement, a landowner with potential red-cockaded woodpecker habitat may have an incentive to harvest timber prematurely and a disincentive to burn forestland, unless woodpeckers colonize the property (Bonnie 1997). Although experimental and no doubt requiring refinement of standards for broader implementation, Safe Harbor Agreements could provide a powerful incentive not only for slowing the loss of species but for actually restoring them.

Safe Harbor Agreements or similarly crafted HCPs could also form the basis for a new habitat mitigation market, as landowners with populations of listed species in excess of the baseline or targeted level set in the agreement "sell" their excess to other landowners who need these credits to offset their proposed incidental takes under an HCP. For instance, in 1999 the International Paper Company initiated the first conservation bank for red-cockaded woodpeckers on its Southlands Experimental Forest in Georgia, where it intends to manage the land for increased woodpecker populations to offset habitat losses elsewhere on its property.

Improving Public Stewardship Programs

Despite the foundation of good landowner assistance programs that state and federal governments provide, their effectiveness is hampered by lack of coordination among agencies and resulting poor delivery to the landowner "clientele." Further, some program requirements set at the federal level hamper successful implementation by the states, or from region to region, owing to lack of flexibility. Finally, even the best programs are starving for the public funding that will enable them to increase their effectiveness—creating a catch-22 of little funding yielding fewer accomplishments yielding less funding. From a structural perspective, government-sponsored stewardship programs generally need to be more convenient for landowners, allowing for simplified "one-stop shopping." They also need to be more fully focused on funding the stewardship activities that landowners want.

Forest Legacy is an example of a New Age federal program dedicated to the conservation of privately owned managed forests. The elements of this program that make it a good model for others include (1) its use of federal, state, local, public, nonprofit, and private partnerships; (2) its leveraging of federal funds with a variety of nonfederal sources; (3) its use of federal policies that set broad guidelines for the goals and functioning of the program, while allowing each state to conduct an Assessment of Need to identify its own priority areas and conservation issues around which to customize its Forest Legacy state program; and (4) its emphasis on the use of conservation easements or reserved interest deeds over fee title acquisition, to maintain forestland in private ownership and management. Further, as part of the overall Forest Stewardship Program, Forest Legacy incorporates permanent protection of private forests with stewardship planning.

Perhaps programs like Forest Legacy do not go far enough in devolving program management to a more local level, with increased relevance to resolving local forest stewardship and conservation challenges. Neil Sampson proposes consolidating all state and federal programs into state forest institutes to provide research, information, analysis, strategic planning, landowner assistance, and public outreach services. Minnesota's Forest Resources Council is a step in that direction.

Organizing for More Public Investment in Private Forests

If people believe that the time has come for society to choose to keep private forests as a significant part of the landscape, then greater investments by the concerned forest community are needed in extensive public education and the organization of effective political coalitions on behalf of private forests. As the data demonstrate, although private forests comprise most of our nation's forestland, there is no guarantee they will remain forested at the dawn of the twenty-second century. Such a public choice cannot be made without investment. The case for conservation of U.S. private forests is being made in certain states and regions, and, periodically at the national level. These efforts need to be expanded to all key forest states and regions, as well as continually on a national basis.

The work that began with the Northern Forest Lands Study and the Governor's Task Force on the Northern Forest Lands in 1988 led to the creation of the Northern Forest Land Council, representing public agencies, forest owners, the forest industry, conservationists, and elected officials across New England. Their 1994 report—and all the research and discussions that surrounded it—elevated private forest issues to a much higher public plane and has spawned many efforts to implement its recommendations. After more than a decade of work, New England is clearly at the forefront of investing in its changing forests. One organization that grew out of this public process is the Northern Forest Alliance, a coalition of conservation, recreation and forestry organizations united to protect the Northern Forest of Maine, New Hampshire, Vermont, and New York. Its 1999 publication *Forestry for the Future* provides a comprehensive and well designed strategic plan for advancing sustainable forestry and forest conservation in the region. The publication clearly describes the issues, the nature of the forests and of the forest-based economy, and how sustainable forestry could be put into practice as part of a sustainable economic plan. Tools, a bibliography, and other resources are provided. More such processes and publications will be crucial to building public investment in private forest stewardship.

Another example of successfully making the public policy case to advance forest conservation is a recent study published by the Society for Protection of New Hampshire Forests and The Nature Conservancy called *New Hampshire Land and Community Heritage at Risk* (Citizens for New Hampshire Land and Community Heritage 1999). Produced by a broad coalition of more than seventy organizations, it uses excellent data to demonstrate the economic importance of open-space values, the costs of development, and the need for greater resource protection. The Trust for Public Land produced a similar publication the same year, compellingly laying out *The Economic Benefits of Parks and Open Space*. Case statements such as these lay the groundwork for a wider political process and, on a national scale, can mobilize a broader base of support.

The seventh American Forest Congress organized through the Yale Forest Forum in 1996 was a good start at engaging a wider representation of the nationwide forest community in finding consensus around a vision for the future of U.S. forests. About 1,200 people attended the four-day conference, surely one of the largest group processes focused on natural resources ever attempted. Although successful in identifying several key areas of agreement, this congress represented only the beginning. Other efforts at building new partnerships of forest interests around common visions and strategies for conservation and stewardship are needed at the local, state, and national levels. Whether these efforts are organized around existing institutions or new institutions are developed, substantially broader forest constituencies need to be assembled to build momentum for wider public engagement and investment in forests. If the various members of the forest community of interest cannot find common ground, political support for improved funding of any of the conservation tools that come out of the public sector will be unlikely to grow.

Seeking support outside the ranks of those who think of themselves as part of the forest community will also be essential. Many forest people are rural, outdoor oriented, older, and white. Most forest resource professionals are white males. Increasingly, the people doing low-wage work such as reforestation and other laborious activities in the southern and western woods are not white, and English may not be their first language. Engaging younger, urban, multiracial, and multiethnic people who may be unfamiliar with the value of maintaining forests is also crucial to the long-term conservation of private forests. Engaging all forest workers equitably as forest ecosystem wealth is restored and maintained will be important too. More broadly, forests need to be better integrated into society. Further, people need to build a society that does not directly or indirectly encourage forest degradation and loss.

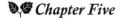 *Chapter Five*

Cultural Tools: Communication, Education, and Assistance

Cultural tools are those that provide forest stakeholders in particular and society in general with the information, options, and motivation to advance private forest conservation and stewardship over the longer term. Through cultural tools, landowners can be better informed about and engaged in protecting their forests for the future. Through cultural tools, forest conservation can become better integrated into forest management, transforming it from simply logging trees to multiresource stewardships. Service providers can become more effective in their stewardship assistance. The wider public can gain a deeper appreciation of the role and value of private forests in their lives. Achievement of all of these objectives is essential to reducing forest fragmentation and loss.

Engaging Landowners in Forest Conservation and Stewardship

Landowner engagement in and commitment to forest conservation and stewardship is essential. Landowners' decisions are primary in determining the future of their forests, whether to increase or decrease fragmentation, timber stocking, or forest investments generally. The presence of more landowners with smaller forest parcels creates a great challenge for communication and engagement. As we have discussed, many of these newer landowners are not very well informed about forests and may view forest management—sometimes even restoration work—in a negative light because they equate it with logging per se. Larger landowners also need to be engaged in stewardship of forests as ecosystems, learning to more effectively manage for biodiversity and other noncommodity values. Each group needs yardsticks to better measure its stewardship performance against some standards. In general, resources and opportunities to engage in forest stewardship have been limited and inconvenient. Further, important conservation and stewardship tools, such as conservation easements and forest certification, are not well understood by many landowners.

Efforts to engage forest landowners in conservation and stewardship—some by old, established organizations and agencies and some by a new generation—are increasing. Some of the programs, organizations, or initiatives described below show how stewardship tools can be better oriented to the character and needs of landowners; how they can be made more accessible and presented in language landowners can understand; and how they can better engage the self-interest of landowners and add value to their lives. When landowners have more understanding of their forests, and are more engaged, they are more likely to actively invest in forest stewardship. Increased stewardship and knowledgeable forest management can bring public benefits such as restoration of degraded resources and reduced fragmentation, and private benefits such as increased earnings—through better stumpage prices, more sustainable harvests, and improved timber quality. However, it is worth noting that not all of these programs address biodiversity directly. To the degree that nontimber values are addressed in many programs, they are focused on issues driven by public policy: fish and wildlife habitat (especially for species of concern or for game species); water quality, and the production of special forest products (mushrooms, ginseng, agroforestry). A few programs are tied to one certification system or another, and several focus on broad ecosystem management.

The Forest Stewardship Program

Several states, such as Pennsylvania, Montana, and Washington, have particularly outstanding Forest Stewardship Programs. Developed by Montana State University extension forester Bob Logan, Montana's program is dedicated to enabling each participating landowner to understand the basics of ecosystem management and how to apply it to his or her own property. In contrast to the usual approach where the government or a consulting forester develops a plan for the landowner, each participant works with program personnel to develop his or her own forest management plan. The program centers on a ten-session landowner workshop series on forest ecosystem management presented by a multidisciplinary team, including field trips and homework, that culminates with each landowner's creation of a Stewardship Plan. Montana State University's publication *Forest Ecosystem Stewardship* was produced jointly with the extension services in Montana, Idaho, Oregon, and Washington, where it is also used. It is a colorful, graphic, and easy-to-understand presentation of the principles and practice of ecologically based forestry, drawing on examples and history from throughout the region. Montana's Stewardship Program shows how to engage landowners' self-interest and love of their forests, teaching them direct stewardship. It uses science, graphics, plain language, and hands-on experience to make a difference.

Master Woodland Manager Programs
Similar to the way in which Montana has crafted its Stewardship Program, some state extension services have created intensive landowner education programs. Known as Master Woodland Manager programs, they aim to both engage landowners in knowledgeable forest management and show them how to share their expertise with their peers. The training program is structured as an eighty-five-hour, ten-session course. It is free of charge in exchange for the graduate's contribution of eighty-five hours of volunteer service educating his or her neighboring landowners. The course includes time spent in the landowners' forests and culminates with each participant's presentation of a plan for his or her own property. As with the Montana Stewardship Program, this course not only prepares attendees to address stewardship issues, it is designed to leverage their knowledge through peer outreach and teaching so an exponential number of landowners can be reached. First developed in Oregon in 1982, the Master Woodland Manager and similar approaches such as Coverts had spread to fourteen states as of 1994, directly training 1,500 landowners (Fletcher and Reed 1996). Coverts was developed in Vermont and Connecticut with the support of the Ruffed Grouse Society. It exchanges training in wildlife management for volunteer teaching of other forest owners. The Vermont Coverts program became so successful, influencing forest management on

52,000 acres, that its graduates formed a nonprofit organization and now direct the program themselves. Ruffed Grouse has since set up similar programs in Massachusetts, Maryland, Ohio, Virginia, Maine, and Minnesota. New York's Master Woodland Owner program and Pennsylvania's Volunteer Initiative Project tie in the Master Woodland Manager–Coverts approach to the Stewardship Program.

The Tree Farm System

Begun in 1941, the American Forest Foundation's Tree Farm System is the oldest national NIPF organization. Its membership includes 70,000 properties covering 25 million acres. Its members are certified according to Tree Farm criteria by one of 8,000 forester volunteers who operate under a state committee. The Tree Farm certification program has evolved from its original goal of promoting forest management planning to one of incorporating general stewardship concepts into plans and activities. Beyond certification, however, Tree Farm seeks to provide continuing education and resources to its members through conferences, workshops, publications, and a Web site. Its Tree Farmer of the Year Awards constitute the most established recognition program for NIPF stewards. Tree Farm has new programs of particular interest. Under its Forests for Watersheds and Wildlife initiative, the organization seeks to network its members with conservation organizations to protect wildlife habitat and watersheds, promote multipartner demonstration projects, and educate landowners and the public. Shared Streams is a program that focuses on protecting and restoring riparian habitat in areas with critical fish habitat, partnering with Trout Unlimited while Forested Flyways seeks to improve waterfowl and neotropical bird habitat in the Mississippi Alluvial Valley, eastern coastal Texas, and the South Carolina coast, partnering with Ducks Unlimited and biologists from International Paper Company.

Forest Owner Associations

There are nonprofit associations of NIPF forest owners in thirty-two states, representing the major commercial forest areas of the United States. These organizations are affiliated through the National Woodland Owners Association (NWOA). Altogether, they count 42,000 landowners as members, holding a median of 82 acres and an average of 600 (Argow, pers. comm. 1999). State forest owner associations vary widely in membership and programs, and usually include the larger forest ownerships that are engaged in regular timber production. These associations typically sponsor conferences and workshops on forest management issues; provide referrals to other forest service providers; publish regular magazines or newsletters; and serve as lobbyists on public policy issues. Field trips to members' forests are a central program of such associations, providing landowners with important stewardship guidance from their peers in the context of particular management situations. The state organizations, such as Forest Landowners of California and the Vermont Woodlands Association, are operated primarily by volunteers. In cooperation with the National Forestry Association and foresters from the Association of Consulting Foresters, NWOA certifies the forest management of its members under the Green Tag Forestry program.

Landowner Assistance Programs
A number of forest products companies have programs to assist NIPFs in forest management planning, effectively providing private-sector forestry assistance to supplement the efforts of extension foresters. These landowner assistance programs (LAPs) seek to increase forest management and timber harvest on forests within the "woodshed" or supply area of a mill. Smurfit-Stone Container, Mead, Georgia-Pacific, and Westvaco have some of the larger programs. Westvaco's is notable for its active assistance in wildlife stewardship activities. The American Forest and Paper Association's Sustainable Forestry Initiative is being used by some of its industrial members to assist NIPF owners in SFI-compliant forest management planning, as a way of ensuring a verified supply of logs from other ownerships.

Landowner Cooperatives
Landowner cooperatives are being created in parts of the Midwest and Northeast to bring together landowners in particular watersheds or locales for mutual education, shared management planning and activities, and often joint marketing to improve returns. Different co-ops have different programs, depending on their history. The co-ops can be informal groups, nonprofit associations, or for-profit organizations. By joining together in some manner, owners of small- and medium-sized forests can potentially gain economies of scale in their forest management, while contributing to a larger landscape level of ecosystem management across their ownerships through joint planning and implementation of activities. They can provide peer support to build forest stewardship. Members can share labor on noncommercial forest

stewardship activities as well as provide mutual information and expertise. Co-op ventures may or may not include joint organization of forester and logging services; ownership of equipment, log-sorting yards, and wood-processing facilities; and joint marketing of log or value-added wood products (or nontimber forest products). Forest management of some co-op forests has been subjected to FSC certification to help set common standards for members—and reassure those members who do not live nearby of the quality of the practices. Conservation easements can also be established across member properties, ensuring continuity in stewardship.

Cooperatives therefore have the potential to help address several barriers to conservation and stewardship: forest stewardship can become more convenient and cost-effective; landowners can gain confidence in forestry activities; smaller landowners can potentially make better returns; and some effects of parcel fragmentation can be ameliorated through joint stewardship.

The Community Forestry Resource Center of the Institute for Agriculture and Trade Policy is seeking to network the various landowner cooperatives. Existing co-ops include the Sustainable Woods Cooperative in Spring Green, Wisconsin; the Hiawatha Sustainable Woods Cooperative in Fountain City, Wisconsin; Vermont Family Forests in Middlebury, Vermont; Wisconsin Family Forests in Hancock, Wisconsin; White Earth Land Recovery Project in White Earth, Minnesota; with others in planning stages in Maine, New York, and Minnesota.

Other Cross-Boundary Stewardship Initiatives

Landowner cooperatives are one example of initiatives seeking to overcome the impacts of parcelization, conflicts over forest management, and the patchwork of multiple ownerships to improve watershed or landscape-level forest stewardship. Some initiatives are focused on conservation of threatened habitat; others encompass broader ecosystem management objectives; and others include local sustainable development. Landscape partnerships provide the opportunity to combine a variety of conservation and stewardship tools focusing on key areas. A number of successful cross-boundary partnerships are described below.

The Wildlife Habitat Improvement Group (WHIG) was organized to improve communication and facilitate peer assistance among NIPF owners in several Vermont towns. The group agreed to consider habitat issues in its forest management. Today the original WHIG area includes 6,000 acres: 2,000 in private ownership and a 4,000-acre state forest. An additional 1,100 acres of state parkland are also in the cooperative. Over thirteen years, WHIG has increased participation in Vermont's special-use property tax program; provided access to state and local agency advisors, as well as to consulting foresters; encouraged joint habitat planning among members; and increased active forest management benefiting wildlife and the local economy. WHIG participants have access to GIS data gathered on their properties, helping them better understand how their properties work together for wildlife. In addition to focusing on the stewardship education of property successors, WHIG participants are now considering how conservation easements can help ensure continuity in forest stewardship for their forests.

Cooperative ecosystem management projects involving public, private, and nonprofit partners include the Applegate Partnership (Oregon); the Mulligan Creek Project (Michigan); the Cannon River Valley Partnership (Minnesota); the Gulf Coastal Plain Ecosystem Partnership (Mississippi); and the ACE Basin Project (South Carolina). As just one example, the Eastern Upper Peninsula Partners in Ecosystem Management (EUPPEM) is an informal collaboration that includes the Michigan Department of Natural Resources, USDA Forest Service, U.S. Fish and Wildlife Service, Mead Corporation, Champion International, and other forest owners in the Two Heart River watershed. This cooperative's goal is to reduce forestry conflicts and facilitate complementary management of forests to maintain and enhance representative ecosystems. EUPPEM has organized the exchange of GIS and other management information among participants; the sharing of data on species and communities; a conference on sustainable forestry in the region; a working group to evaluate and advise on member management practices; guidelines for management of bird species of concern; and public education regarding the effects of deer overpopulation on the forest.

The Monadnock Landscape Partnership (New Hampshire and Massachusetts) and the North Quabbin Regional Landscape Partnership (Massachusetts) are two collaborative land conservation and stewardship efforts in New England forest areas that bring together government agencies and nonprofit conservation organizations with the goal of sharing information, public education, and land/resource protection projects.

Land Trusts and Conservation Easements

Landowners are increasingly turning to land trusts as partners in meeting their conservation objectives. According to the Land Trust Alliance (LTA), the national umbrella organization for nonprofit land conservation groups, there are about 1,225 land trusts (or conservancies) in the United States—an increase of 63% since 1988 (LTA 1998). Collectively they had protected 4.7 million acres of land, including about 1 million that have been transferred into public ownership, as of 1998. Although there are several well-known national organizations such as the Trust for Public Land, The Nature Conservancy, the Conservation Fund, and a few relatively large regional ones, most land trusts are smaller, local entities. Being local, state, or regional organizations, land trusts can reach out knowledgeably to landowners in their areas to assist them in stewardship issues. In forest regions, land trusts can assist private owners on a private, nongovernmental basis in conserving their property to meet their economic and stewardship objectives while ensuring that the forest cannot be converted to other uses in the future.

While land trusts often acquire fee title to property to create ecological preserves or for public parkland, the primary tool used by forest landowners working with land trusts is a conservation easement (CE). Conservation easements have been in use since the 1930s, with expanded application since their tax deductibility was confirmed by the IRS in the 1970s. In 1998 LTA found about 7,000 conservation easements in existence covering 1.4 million acres, an increase of 378% from the 290,000 acres of CEs in 1988. Of these, an estimated 70% cover forestland (LTA 1998). Since that survey, the use of conservation easements has been expanding widely.

Historically, CEs were used primarily to protect scenic vistas, agricultural land, or threatened habitats on private lands. More recently they are being applied to forests where productive management continues on the property. CEs are partial interests in property voluntarily granted by a landowner to a land trust (or governmental agency) to restrict certain land uses, thereby permanently protecting the forest from conversion or damaging management practices. CEs are site specific to the property and are therefore more flexible in meeting public benefit objectives in the context of private ownership. The title to the property stays with the landowner and it can be sold or transferred like any other property, subject to the terms of the CE. The restrictions in the CE address nonforest development (such as subdivision and residential building) as well as forest management activities that can negatively affect environmental values on the property. The land trust holds these partial interests or restrictions in trust ensuring that the terms of the CE established by the landowner are not violated through time by monitoring and, if necessary, legal enforcement. CEs can be charitably donated or sold to the grantee, with the value in taxes saved or revenue received providing a significant incentive for conservation. The landowner is thereby compensated for his or her commitment to conservation while the public obtains a lasting benefit for its investment.

Well grounded in the law and application, conservation easements provide the best available method for landowners to permanently prevent fragmentation and conversion of larger properties while keeping them in private ownership and productive use. CEs also ensure the continuity of management, protect landowner investments in stewardship from future liquidation, and provide economic returns to landowners for protection of forest values. They can be combined with forest certification standards, and incorporate various habitat restoration and other stewardship activities. Despite these appealing qualities, however, the application and benefits of CEs for private, managed forests is still not well understood by landowners and their advisors, pointing out the need for more extensive and targeted educational initiatives for this tool.

Sometimes, as an alternative to permanently restricting development on their property, landowners enter into cooperative management agreements with land trusts, which then assist in identifying important habitats or other conservation values and providing stewardship advice. For instance, The Nature Conservancy recently began helping Westvaco Corporation to survey a million acres of the industrial owner's forests in the South to help it identify and protect key habitats within its forest management planning. The Nature Conservancy also has a comanagement agreement with Georgia-Pacific for 22,000 acres of company forestland on the ecologically sensitive Lower Roanoke River in North Carolina.

In addition, land trusts provide public and landowner education on conservation while some are growing to provide stewardship assistance generally. As community organizations dedicated to conservation, land trusts can reach out to many forest owners who find traditional landowner organizations not sufficiently stewardship oriented. Topics of landowner workshops include road building and maintenance, estate planning, salmon restoration, and erosion control, among others. Some, such as The Nature Conservancy, Tall Timbers, and the Pacific Forest Trust (PFI), engage in research and demonstration projects related to ecological forest management.

A small but growing number of land trusts specialize in the challenges of private forest conservation and stewardship, recognizing that forests are both complex, natural ecosystems and often working assets for landowners.

These land trusts include the Forest Society of Maine, the Society for Protection of New Hampshire Forests, the Vermont Land Trust, the Adirondack Land Trust/The Nature Conservancy (New York), the New England Forestry Foundation, Tall Timbers Research Station/Red Hills Conservation Program (Florida-Georgia), certain state chapters of The Nature Conservancy (for instance, Maine, Vermont, Minnesota, Georgia, Virginia), and PFT (California-Oregon-Washington). Given the complexity of CEs for managed forests as well as their long-term monitoring demands, land trust forestry knowledge and organizational capacity needs to be further developed to expand this service to landowners in many other forest states.

Stewardship Forestry Organizations

There are a variety of local and regional nonprofit organizations dedicated to sustainable forestry. In their areas, they often provide valuable landowner education and assistance. Like land trusts, these organizations often function on a grassroots level, close to local landowners; they also can appeal to landowners who are not active in traditional forestry associations. The Institute for Sustainable Forestry (ISF) in California is one such group. ISF provides landowner workshops on forest restoration, forest management planning, forest certification, and other topics, often in partnership with government agencies or extension agencies. The Mountain Association for Community Economic Development (MACED) is a sustainable development nonprofit organization based in Kentucky, focusing on the Appalachian region. Among other services, MACED is producing user-friendly information to help landowners understand and manage their forests more sustainably. These efforts include an interactive handbook that guides landowners through the process of developing a management plan for their forests. MACED's materials help landowners better understand their options for management and forest succession (estate planning). MACED is also developing demonstration forests for public and landowner education as well as seeking to catalyze the development of land trusts to provide forest conservation services in the region.

Helping Foresters and Loggers Provide Stewardship Assistance to Landowners

Foresters and loggers are often the primary sources of information used by landowners in making decisions about forest management. Although many provide stewardship assistance, unfortunately others do not. When foresters and loggers more fully integrate forest ecology and conservation into their management services, they will not only benefit landowners and forests but also will improve their own businesses. This is because NIPFs have indicated in numerous surveys that they are more interested in timber harvesting when it is in the context of overall forest stewardship. By adopting a stewardship-oriented approach— and a longer-term perspective—to their professions, foresters and loggers can provide ongoing services rather than those simply focused on a particular commercial harvest. Engaging loggers' and foresters' self-interest to advance forest stewardship and conservation requires education and incentives.

Increasingly, undergraduate and graduate forestry programs are recognizing the importance of producing *forest stewards,* not simply timber specialists. Increasingly, colleges are integrating their forestry programs within schools of natural resources or environmental studies. The Sustainable Forestry Partnership is an example of a joint effort among forestry schools, led by Oregon State University and Pennsylvania State University, to develop new curricula more fully incorporating stewardship. Ongoing forester education is more important now that the science of forest ecosystems is growing daily. The Society of American Foresters, the Association of Consulting Foresters, and other forester groups can play greater roles in changing the paradigm of practicing foresters from a focus on timber growth and yield (with protection of the forest ecosystem as a constraint) to one of enhancing ecosystem productivity (with high-quality timber and other products as expressions of a healthy forest). The Forest Stewards Guild is a new professional organization founded in 1997 to "promote ecologically responsible resource management that sustains the entire forest across the landscape."

There are an estimated 37,000 logging firms in the United States, employing some 195,000 people (AF&PA 1999). Training for loggers has not been as extensive as that for foresters. Yet it is the logging operator who leaves the most lasting imprint on a forest during timber harvest. Peer-based logger training is growing, incorporating not only techniques of operation and equipment safety but also forest ecosystem science and silviculture. Logger Education to Advance Professionalism (LEAP) is a recently developed extension program in use in twelve states as a pilot effort to train loggers as applied ecologists. Based on a highly successful training program in Vermont, in which half the state's loggers graduated within five years, LEAP improves logger communication with foresters and landowners, reduces water quality and other forest practice violations, and improves the quality of residual stands. The Game of Logging® is a peer-training program based on competition that has been widely used in the forest products industry. An important element of the AF&PA Sustainable Forestry Initiative (SFI) is the requirement for training and certification of logging crews in BMP and other environmental standards.

Advancing Understanding of Stewardship Forestry

Forestry based on ecological principles is still young. We have much to learn about the use of silviculture and other practices to restore and conserve forest ecosystems generally and biodiversity in particular. Public funds directed toward the research programs of the USDA Forest Service, as well as through land grant universities, are fueling great advances in our understanding. However, communication and application of this knowledge are lagging. In particular, demonstration and quantification of the costs and benefits of stewardship-oriented forestry are in short supply— unknown, in fact, to many landowners, resource managers, and the public. Researching, packaging, and distributing better, more scientifically accurate forestry stewardship information for users through a variety of media require more investment than is currently available. For instance, the partnership among the U.S. Forest Service Pacific Northwest Research Station, the Oregon State University College of Forestry, and the university extension service in Corvallis, Oregon, allows for an integration of basic research, teaching, demonstration projects, field trips, workshops, and publishing that can hasten the introduction of new forest stewardship knowledge into application. Also toward this end are several excellent publications that have been written for landowners and resource managers in the last few years, including *Forest Ecosystem Stewardship* (described above), *Creating a Forestry for the 21st Century* (Kohm and Franklin 1997), and *Good Forestry in the Granite State: Recommended Voluntary Forest Management Practices for New Hampshire* (New Hampshire Forest Sustainability Standards Work Team 1999). Demonstration forests, whether operated by private owners, universities, or conservation organizations, can provide not only important data for education but also very high-quality on-the-ground educational opportunities. Other efforts worth noting and emulating are described below.

The Minnesota Forest Bird Diversity Initiative
Initiated in 1992 by the Minnesota Department of Natural Resources, this project was aimed at developing landscape management tools to maintain Minnesota's rich diversity of forest birds. The project includes expanding forest bird monitoring; conducting bird productivity studies; modeling relationships of forest birds to landscape characteristics; and promoting forest bird conservation and management. A computer model, LANDIS, has been developed to help forest managers understand how bird populations would respond to changes in forest cover types and landscape vegetation patterns as the result of logging, land use change, and natural disturbances. In 1995, project staff published an award-winning book, *Birds and Forests: A Forest Management and Conservation Guide,* which was distributed free of charge to 1,725 forest managers throughout Minnesota. Project staff were also integral to the development of five logger education workshops in 1996 that reached six hundred loggers. That year they also made thirty-three presentations to local, regional, and national audiences. Since that time the project has also yielded a forest stewardship publication, *Planning for Birds,* as well as a series of FSP workshops called "Birds and Forests."

The Washington Forest Landscape Management Project
A joint effort begun in 1992 by the USDA Forest Service, the Washington Department of Fish and Game, and the Washington State Department of Natural Resources, this project investigated the feasibility of integrating forest management across ownerships to "increase the likelihood of sustaining viable populations of sensitive wildlife species and fish stocks, while reducing landowner costs and uncertainties" (Carey and Elliott 1994). The project involved (1) developing and evaluating scientifically credible landscape management techniques to conserve fish and wildlife; (2) providing information on the development and maintenance of managed forests that furnish habitat for wildlife and fish species sensitive to forest management; (3) determining methods for producing quality wood at an acceptable rate of return to landowners; (4) identifying pragmatic incentives for landowners to participate in landscape management; and (5) analyzing landscape management alternatives for each degree of

species protection along with economic costs and benefits. Innovative project staff convened a multidisciplinary scientific team and have produced several publications on their research findings for all landowners and managers seeking to manage forests for ecological and economic values.

Stewardship Forestry at Work

A PFT project, this is a series of case studies of nonindustrial landowners in Oregon, Washington, and California whose excellent forestry practices are improving forest health, biodiversity, and production of high-quality wood products. Each case preparation included fieldwork by PFT foresters and meetings with landowners. The resulting publication will be graphically appealing and written in language accessible to other forest landowners, allowing for peer sharing of best stewardship practices for different forest types with different stand conditions, and including financial analyses of costs and returns. *Stewardship Forestry at Work* will be distributed to forest landowners and managers in the region. Its results will also be promoted through the general media to provide the public with images of how good forestry benefits ecosystems.

Advancing Understanding of Forests and Forest Owners

All forest stakeholders need more, better, and more timely information on our private forests, including condition, threats, and other factors. Existing information also needs to be more accessible and useful to owners, managers, and the public. Better understanding of particular forest ecosystems can improve their conservation. By providing better information, we can produce more relevant analyses of it, which will allow more effective responses to pressing issues. And we can begin to ask better questions, so future information is much more useful than today's.

Much of the data that exist are not well integrated. At the subregional and local levels, the data sets are especially lacking in useful information, having been produced for larger scales. Fortunately, the USDA Forest Service, USGS, NRCS, and other key federal agencies are moving to provide databases and publications over the Internet and initiating more frequent data collection, as with the conversion of the FIA to a continuous inventory system. State resource agencies are collecting, organizing, and providing more resource data on-line. The Montreal Process is seeking to create internationally accepted scientific indicators for sustainable forest management. As we have discussed elsewhere in this book, much remains to be accomplished. The situation regarding our understanding of private forest owners is similar. Following are just a few of the organizations and initiatives working to address this problem.

As described in chapter 2, the Southern Forest Resource Assessment is the first attempt to integrate the data and analyses of all relevant state and federal agencies to better understand the condition of this important region's forests. If successful, this model should be followed nationwide. A less comprehensive but regionally important effort was the interagency production of the Southern Appalachian Assessment, organized by the Southern Appalachian Man and Biosphere Program in 1996.

Geographic Information Systems
GIS mapping of threatened forests at the national, regional, and state levels can be a powerful tool to focus public and private forest conservation efforts. Some strategic mapping approaches are described below.

NIPF Population Pressure Index
As part of the research for this book, The Sampson Group (TSG) used GIS to create a map that illustrates areas where population densities are heaviest and where they exist in proximity to private forestlands, highlighting priority areas based on an NIPF Population Pressure Index TSG created. The group used data mostly available on the World Wide Web, including county-level population estimates for 1990 and 1998, to indicate recent population growth rates. These rates were then used to compare the fast-growth areas to land use and ownership estimates for 1992. Population statistics were downloaded from the U.S. Census Bureau; forestland statistics and ownership were obtained from the USDA Forest Service; and county-level nonfederal land estimates were obtained from the NRCS.

Using county-level data obviously introduced some errors, since the forest inventory information becomes less accurate below the state level. The use of these data also introduced a level of generality that must be considered, since all the growth may occur in one corner of the county instead of being spread across the landscape. These were, however, the best available data sources. TSG took one additional step that makes these data more specific, calculating population density as the number of persons per 1,000 acres of private land. Such a calculation makes little difference in eastern areas, where the majority of counties contain little or no public land. In the West and in the more rural eastern counties, however, it can make a great difference. While nearby populations place recreational pressures on public lands, conversion and fragmentation pressures are focused on private lands.

TSG then extracted the acres of nonfederal land by county from the National Resource Inventory (USDA-NRCS 1994), seeing this as the best available basis for estimating private lands at the county level. However, the process also introduced errors in counties where the nonfederal total includes state or county government lands. Nevertheless, there are no readily accessible data sources with which to correct these errors, and we felt them to be minor in relation to the total. Estimating the 1998 population density as a function of people per nonfederal acre gave one indicator of population pressures on those lands. One additional step was needed, however, to separate those areas where population density negatively affects primarily agricultural lands or grasslands: TSG divided the estimated acreage of NIPF forests per county from the 1992 FIA database by the amount of nonfederal land per county. This produced a fraction representing the proportion of nonfederal land that contains NIPF forests. Multiplying that fraction by the number of people per 1,000 nonfederal acres produced the NIPF Population Pressure Index, which reflects how a county's population density is likely to affect NIPF forestlands. The Virginia data reported by Wear (1999) on the decreasing probability of forestry management as population density increases were then utilized to give a low- to high-population-pressure index range.

Bioregional Information

A bioregional information system for the North American rain forest, Inforain (a project of Portland, Oregon–based Ecotrust), makes available over the Internet GIS databases compiled from public resource management agencies. The databases provide regional and community information to support conservation-based resource management and economic development. Inforain also produces bioregional and watershed-level atlases.

Metro-Area Analyses

Using a methodology similar to that utilized by DeForest et al. (1991) in their study of southern statistical metropolitan areas (SMAs), mentioned in chapter 2, researchers can generate local, state, and regional analyses of the likely impacts of expanding urbanization on private forests. Metropolitan and nonmetropolitan counties can be selected using USDA Economic Research Service urban-rural continuum codes to classify counties by population. Population densities (people per acre, or ppa) can be estimated using census data. Future trends in population growth can be inferred from census data and other government projections. USFS FIA data can provide county-level information on acreage of timberland, standing sawtimber volume, growing stock volume, timber growth, and removal. Overlaying this information can provide insight into the urbanization of private forestland in selected areas, as well as into future trends of development in forest areas.

Integrating Forest Ecosystem Data
GIS tools are essential to integrate the types of population growth and urbanization indices described in previous sections with information on the status of actual forest conditions.

Using a variety of data sets available from state and federal agencies, nongovernment organizations, and satellite imagery, researchers can determine the structural characteristics of forest landscapes such as percentage of cover, amount of disturbance, fragmentation, and patch size. Several types of remote sensing data have been used on a local (county-level) scale to assess forest health related to insect infestation, acid rain, sedimentation, and other variables. Databases of particular value on vegetation condition and biodiversity at local and bioregional scales include USGS GAP data for many states as well as databases from The Nature Conservancy, the World Wildlife Fund, the Wilderness Society, and Ecotrust.

The Wilderness Society's Center for Landscape Analysis in Seattle used satellite imagery to assess forests at risk in the greater I-90 highway corridor in the Cascade Mountains to facilitate land exchanges (*www.tws.org/ccc/pacificnw/cascade.htm*). Ecotrust examined the condition of forested watersheds from San Francisco, California, to Kodiak Island, Alaska, to determine an index of forests at risk based on cover, percentage of disturbance, and patch size. Ecotrust has also used satellite imagery to aid understanding of differences in forest harvest rates for private and public lands in southeast Alaska (*www.inforain.org*). The Woods Hole Research Center has recently completed an assessment of land use change and forest loss for all fifteen towns of Cape Cod, Massachusetts, over a forty-year period (1951 to 1990) (*www.whrc.org/ccatlas/ccatlas.htm*). The

Pingree Forest Partnership in northern Maine is preparing a GIS and forest condition database for the completion and monitoring of a conservation easement for 754,000 acres of productive forestland (*www.neforestry.org*). Many watershed organizations around the country have developed GIS databases to facilitate their understanding of the status of forestland conditions and ownership.

User-Friendly Scientific Publications

The Redwood Forest: History, Ecology and Conservation of the Coast Redwoods is a book produced by the Save-the-Redwoods League that provides the first accessible, scientific overview of this significant and threatened forest type. A consortium of foresters and scientists provide key information about the redwoods in a readable format for forest managers, conservationists, and the general public. Whether or not one agrees with the conservation strategy advanced in this book, all major forest types should have such useful information available for applied conservation.

Two southern organizations are undertaking similar work to save the vanishing longleaf pine forests. Based in Alabama, the Longleaf Alliance promotes the ecological and economic values of longleaf ecosystems, seeking to restore and enhance this highly threatened forest type. It works cooperatively with educators, landowners, conservation groups, and forestry organizations to disseminate practical information about longleaf forest management. In an affiliated effort, the Sandhills Area Land Trust published a colorful, instructive booklet, *A Working Forest: A Landowner's Guide for Growing Longleaf Pine in the Carolina Sandhills.*

Reconnecting People and Forests

A broad section of the urbanized American public needs to be reconnected with forests—not just with forests as wilderness, but with private, managed forests. Accomplishing this could greatly accelerate the conservation of forests, because public interest and support will grow to make the necessary investments. Further, considering the 93% of forest owners with less than 100 acres, they *are* the general public in most respects—and engaging them in the conservation of their own forests may only be accomplished through a strategy targeting the wider public. Improving the reputation of forestry among the public will be an essential part of this effort. Promoting the role of private forests in providing desirable public benefits will be central to any such campaign. Landowner stewardship of nontimber values—such as water quality, wildlife, and recreation—must be emphasized.

New messages and media are needed for new audiences. Wider use of general media, expanded educational programs at all levels, more recreational opportunities in private forests, and public participation in stewardship activities can all play a role. Creating and supporting a carefully crafted public education campaign—really a marketing campaign—will require a coalition representing not just the forest products industry or conservationists, but a suite of stakeholders representing the facets of private forest values. It will likely also need to be pursued at state, regional, and national scales.

Current initiatives in this regard are limited in number, indicative of the active forest community's focus on more immediate issues. A few are noted below.

New Educational Outreach to K-12
Like so many stewardship tools, the new generation of educational programs for forests needs to be grounded in ecological understanding and not simply in the romance of logging or the love of pristine wilderness. *See the Forests* is a Vermont program organized by Shauna Ratner of Yellowood to take kids out into the woodlands of their communities so they can learn how their management provides all kinds of things they enjoy, from wood products to wildlife. It is particularly important to involve youth—especially urban and low-income youth—in actual hands-on stewardship projects. *Adopt-a-Watershed* is a program that gets kids involved in the dynamics of their local ecosystems. Similar efforts can be undertaken by private-sector partners in cooperation with public programs, such as California's Outdoor Education Program or the California Conservation Corps, which hires disadvantaged youth to work on restoration projects. The Sierra Club has started such a program called *Youth in Wilderness.* Why not *Adopt-a-Forest?* It is important to remember that publications such as *Ranger Rick,* from the National Wildlife Federation, can have a great impact on teaching the value of managed forests.

Backyard Habitat
Teaching folks that ecosystem management begins at home, even on small properties, helps bridge the gap to understanding larger scales. In fact the

techniques used by the two programs mentioned here could be readily incorporated into an outreach strategy for smaller NIPFs in the United States. The National Wildlife Federation's Backyard Wildlife Habitat program is one such effort. Using a partnership network, it incorporates volunteer training, peer mentoring, networking of trained volunteers, and promotion of habitat projects in participants' areas. Backyard Conservation is a campaign organized by NRCS in partnership with the National Association of Conservation Districts and the Wildlife Habitat Council to promote natural resource conservation practices on nonagricultural private lands, from city backyards to country properties. To facilitate participation the program has a Web site from which tip sheets of recommended practices can be downloaded; links to other relevant programs and resources through the Web site; and a toll-free phone line. NWOA has recently initiated a similar Backyard Woodlands program with NRCS.

Demonstration Forests

A traditional tool for both public and landowner education, new networks of demonstration stewardship forests need to be organized and marketed to new audiences through new partnerships. For instance, Oregon State University's extension service utilizes field trips as central elements of its workshops on ecosystem management and new silviculture. The Society for Protection of New Hampshire Forests and the New England Forestry Foundation both own forestland that they manage for stewardship education as well as revenue. State forest owners' associations organize field trips to members' forests that are open to the public as well. Owners of managed forests in the urban-rural interface who provide public education opportunities on their lands can help foster better understanding of the benefit of forest stewardship management.

Improving Public Confidence in Forestry through Certification

Forest certification is a new and growing movement that seeks to set high environmental standards of forest management performance to build public and specifically consumer confidence in forestry. Forest certification can also provide NIPFs—especially those who don't live on their property—with assurance on the quality of forest management occurring on their land. Various certification systems, such as the Forest Stewardship Council (FSC), the American Forest & Paper Association's Sustainable Forestry Initiative (SFI), NWOA and the National Forestry Association's Green

Tag Program and Tree Farm, are similar in their broad standards but vary in the specifics. FSC is the most focused on its inclusion of ecosystem values and the most detailed in its performance standards. FSC was the first to require third-party—as opposed to self- or interested-party—certification based on an on-the-ground review of actual practices by its expert teams. FSC also provides certification of the chain of custody so consumers can buy wood products identified as sustainably harvested with greater assurance.

All forest certification systems can help engage forest owners and their managers in better understanding and investing in forest resources and stewardship challenges. Each system requires the creation of a forest management plan according to its guidelines. A publication of the Pinchot Institute for Conservation titled *Understanding Forest Certification: Answers to Key Questions* is a useful introduction to certification approaches currently in use. The Society of American Foresters has also produced a helpful overview of certification.

Both SFI and FSC are engaged in outreach campaigns to build public confidence in the integrity of their forest certification systems as part of their branding efforts for participants. Such promotion efforts can play a very important role in reassuring the public that not all forestry is bad—and that it is worthwhile to support "the good guys." Such efforts can dovetail with other public engagement efforts.

🍃 *Chapter Six*

Financial Mechanisms and Markets for Conservation

It is difficult for conservation to compete economically with conversion. Financial markets are focused on short-term returns. In the case of private forests, this primarily means marketing timber or land development. Therefore, if more private forests are to be conserved, conservation-based financial instruments and other market-oriented incentives need to be expanded or developed. This is especially true if we are to engage larger financially motivated private landowners in forest conservation. These ownerships are managed to provide profits to investors—whether holders of publicly traded stock or investors in institutional pension funds. More of those forests will be conserved for public benefits when their owners can gain some combination of added revenue, reduced costs, improved community relations, and regulatory stability.

A number of current and emerging initiatives aim to better align capital markets with forest ecosystem characteristics and generate financial returns from stewardship of ecosystem assets. These initiatives could be especially powerful in expanding the direct conservation of threatened forests in the near term. In addition, they can fuel the success of a new stewardship forestry business model in which management for conservation of biodiversity is a competitive advantage. This section of the toolbox therefore seeks to accomplish conservation

165

through a combination of market interventions and the "build-a-better-mousetrap" strategy.

Improving Returns from Stewardship Timber Harvests

In a world that is recognizing that wood is not a limitless resource but is in fact an increasingly scarce and highly valuable asset, maximizing the value of trees that are harvested is an important strategy to increase returns from stewardship-oriented management. There are various ways to accomplish this objective.

Expanded Branding for "Good Wood"

Wood produced using excellent forest practices can be branded under a certification system such as FSC to build market recognition for the producer's commitment to stewardship. Market recognition can take the form of developing or expanding market share in a defensible niche, and/or obtaining a premium price over nonbranded wood. As public recognition of the need for consumer support of forest stewardship grows, so can the market for certified wood.

Development of Local Value-Added Wood Markets

Many forest regions lack market opportunities for higher-return, value-added wood products, thereby limiting returns from timber harvest. For instance, for lack of markets and good forestry information a landowner may sell his or her hardwoods for pulp rather than for sawlogs. Or, conversely, a landowner may need to do restoration thinning of an overstocked and high-graded hardwood forest but lack a market for these undervalued logs. Smaller wood harvests can yield greater returns to landowners and communities through development of local value-added enterprises. Landowner co-ops can invest collectively in value-added milling equipment, enabling them to triple the stumpage value to their members by capturing that value-added increment themselves. States such as Michigan realized some time ago that by keeping value-added processing and finished products manufacturing in local wood-producing communities, they could generate up to thirty-five times the raw wood value.

Expansion of Business Models Based on Mature, High-Quality Timber

Mature, large-dimension sawtimber is already scarce in the United States and economic trends clearly predict its obsolescence in the near term. As older trees have become more scarce, the mainstream of the forest products industry has created substitutes through engineering wood products out of chips and smaller-dimension lumber. However, as ecological science demonstrates, these bigger, older trees are essential components to fully functional forest ecosystems. Therefore, the challenge is how to restore older forests on the landscape and make it worth the landowner's investment of time and money to do so. The answer needs to include full-cost pricing for older trees, providing a premium to landowners for that extra commitment to grow and maintain mature forests. Although market opportunities and pricing for older hardwoods have increased as supplies have become more scarce, equivalent opportunities for softwoods have not. Large-dimension logs from both hardwood and softwood trees have attributes that lend themselves to high-margin products, leaving smaller and lower-quality trees for lower-end commodities. However, this change in product mix is recent and opportunities for better utilizing high-quality timber need to be redeveloped for a new era of scarcity. This is an area of strategic market development that lends itself to private and nonprofit partnerships.

Expanding Nontimber Revenue Opportunities

A variety of existing and potentially profitable forest products can be produced through conservation-based management to enhance returns from forest stewardship. Some of these products are described below.

Market Development for Special Forest Products

Using various partners, local and regional markets can be expanded for sustainably harvested nontimber forest products such as edibles, floral greenery, medicinals, and other items. Drawn from the great diversity of forest plant life, special forest products have the potential to provide forest owners with regular, annual income that can finance forest carrying costs and fund further stewardship. Leasing to harvesters (subject to a good contract with harvest guidelines) can yield returns in the range of $5 to $15 per acre and owners who invest in primary processing have the potential to double or triple those returns. Some market development activities have been undertaken by the USDA Forest Service and university extension services. One nonprofit/private-sector collaboration is the creation of Rainkist by the nonprofit Shorebank Pacific Enterprises, a subsidiary of Shorebank Pacific (itself a partnership between Ecotrust and the Shorebank Corporation). Based in Olympia, Washington, Rainkist is a marketing organization that represents twenty-five small processors and manufacturers of special forest products gathered in the temperate coastal rain forest of the Pacific Northwest. Incubated within the nonprofit business development program of Shorebank Pacific Enterprises, Rainkist is being spun off as a for-profit affiliate. Rainkist provides business advice, microloans, retail and wholesale market representation, and forest stewardship training for its members to help ensure that the harvesting is conducted in a sustainable fashion. In fact Rainkist has produced a practical manual of best practices for gatherers called *Specialty Forest Products: Stewardship Field Guide.*

Developing Private Forest Recreation Enterprises

Well-managed private recreation also has significant potential to increase returns in a way that funds forest stewardship. The use of forests for hunting, recreation, and personal renewal is age old. With increasing population, expanding urbanization, increasing disposable income, and crowding of parklands, fee-based private forest recreation is a growing revenue source. Consumptive (hunting and fishing) and nonconsumptive (hiking, bird-watching, horseback riding, biking, photography) recreation can be compatible with other forest stewardship goals, including timber harvest activities. Nature-based tourism in particular is growing at a high rate and the USDA Forest Service estimates that accessible private forests can fill a growing market gap in the provision of quality recreational opportunities in the decades ahead. Investments in habitat enhancement in prime hunting areas (for instance, quail hunting in the remaining longleaf pine forests of southern Georgia and wetlands restoration in bottomland forests of the Mississippi Valley) can dramatically increase recreational fee income while maintaining or restoring complex natural forest ecosystems. Depending on the hunting market and habitat quality of the forest, landowners can earn $2 to $15 per acre for a season. On the best-quality bird-hunting forest properties in the Southeast, fees can be two to three times higher. Although several large landowners such as Champion International and Anderson-Tully have conducted successful hunting and recreation programs, NIPFs in certain regions do as well. In general, successful operations have shown it is possible to add 15 to 20% to the net present value of forest investments through compatible recreational development. Opportunities exist for private forest owners to receive assistance in habitat enhancement from wildlife organizations such as Ducks Unlimited, Trout Unlimited, and the Rocky Mountain Elk Foundation.

Developing Markets for Ecosystem Services

The capacity of forests to provide immensely valuable ecosystem services is well documented at the macro level. Costanza et al. estimated the global economic contribution of ecosystem services of all kinds at $33 trillion, of which $4.7 trillion is attributable to forests (1997). Primary among these are contributions of fresh, high-quality water supplies for habitat, human consumption, and hydropower; carbon sequestration for climate stabilization; pollination and biocontrols for agriculture; banks of genetic diversity; waste treatment; and flood and storm protection. Although ecosystem services may form the foundation of much economic activity, forest owners have realized very limited monetary returns from managing their forests to yield these services. Poor forest management that degrades the quality of ecosystem services may increase economic and social costs, but good forest management that enhances the capacity of forests has not generated revenue because there has been little market development. Scarcity and social need are driving the emergence of markets for carbon sequestration and watershed services. New institutional mechanisms are being developed to finance more forest conservation and stewardship to optimize the provision of these services. The potential for mobilizing new capital into private forest conservation by developing markets for these ecosystem services is considerable.

Forest-Based Carbon Sequestration
Forest carbon stores can be increased to greatly reduce atmospheric CO_2, a key global warming gas. Protection of forests is a primary goal identified in Article 2 of the Kyoto Protocol for improving the world's carbon budget. Continuing negotiations for international cooperation on climate change are expected to provide the opportunity for carbon producers to claim some form of emissions reduction credits from forest conservation, improved forest management, and reforestation not only in developing countries but in the United States as well. Generally accepted accounting procedures for measuring and monitoring forest carbon stores are still developing; nevertheless, it is agreed that there are four basic ways to increase net forest carbon stores over a baseline of business as usual.

The first way is to protect forests threatened with development from conversion and protect remaining old-growth forests from harvest to maintain these existing major stores. Second, reforesting previously cleared forest areas such as agricultural lands will increase net forest carbon stores. Third, growing managed forests to older ages and retaining portions of the stand after logging can result in more carbon accumulation before and after harvest. Finally, altering harvest methods and site preparations for replanting to reduce impacts to soil carbon will increase net forest carbon stores.

Although virgin tropical forests retain great stores of carbon, managed temperate forests contribute to increasing carbon stores as well. U.S. coastal temperate rain forests in fact excel in their biological capacity to store biomass, accumulating two to four times the carbon per acre of tropical forests. These species, primarily redwood and Douglas-fir, are known to store the greatest carbon tonnage per acre of any forest type (Turner et al. 1995). U.S. private forests have other competitive advantages as well, including the depth of knowledge of their ecological dynamics, the security of legal tenure, and the country's relative political stability.

The size of the market for forest carbon reduction credits is very difficult to quantify, as are the prices. The World Bank estimates that world demand for carbon offsets available through flexible market mechanisms may amount to 500 million tons of carbon annually during the first budget period under the Kyoto Protocol (2008–2012). Existing forest-based carbon projects, as well as projections by the World Bank, suggest a price range of $5 to $30 per ton of carbon stored (or an average of $17.50 per ton). Given the early stage of the market and the limited number of projects, no true market pricing has yet developed. If the United States were to seek to offset 20% of its 1997 emissions through increases in domestic forest-based stores, that would create a $5.25 billion annual market.

A 1999 analysis by PFT of changes in on-site forest carbon stores resulting from changes in silviculture in MacMillan Bloedel's forests in British Columbia (now owned by Weyerhaeuser) shows that at a price of $10 per ton of additional carbon stored, the sale of carbon reduction credits by the company for a ten-year period could yield more than $53 million. Based on stumpage rates prevailing at the time of the analysis for Douglas-fir and hemlock, the dominant species in these stands, this amount exceeds foregone timber harvest income in the same period (Pacific Forest Trust 1999).

As well-established and permanent legal mechanisms to conserve forests and guide long-term forest management, conservation easements (CEs) are appropriate means to secure tradeable forest-based carbon reduction credits on particular forest properties in the United States and other countries where CEs are utilized. The value of CEs is determined by the opportunity cost of foregoing conversion and nonforest development, as well as the cost of restricting forest management options to enhance carbon stores, for example, ensuring prompt reforestation after harvest, extending rotations, managing portions of the property for protection or restoration of old-growth stands, and making other changes in silviculture and terms as may be appropriate to the site. As demonstrated by the modeling and pricing of potential carbon credits for MacMillan Bloedel, the sale of the credits could pay the full cost of conservation.

Now is the time that market-making investments in high-quality forest-based carbon emissions-reduction projects can most strongly influence the development of this new market. Whether produced in the United States or abroad, such projects need to use credible scientific standards, with transparent accounting and ongoing third-party monitoring; and forest project carbon credits need to be based on management actions that are demonstrably additional to current regulatory baselines. As with any early-stage market, discontinuities in supply and demand need to be overcome to expedite the development of a functioning marketplace. Public, private, and philanthropic entities interested in the potential of the forest-based carbon market can influence its development by investing at this early stage of market development through supporting programs such as those discussed below.

PFT is currently the only market maker in U.S.-based carbon emissions-reduction projects based on forest conservation and stewardship. PFT is acquiring CEs through its *Forests Forever Fund* in which carbon emissions reduction credits based on the conservation accomplished are transferred to PFT, where they are inventoried for resale to fund further conservation. In 2000, PFT made the first sale ever of carbon credits based on forest conservation in the United States to Green Mountain Energy, a power supplier. Using charitably donated or low-cost carbon credits and philanthropic capital, the *Forests Forever Fund* functions as a carbon bank to facilitate Pacific Northwest forest-based transactions. Resource Conservation and Development Councils (nonprofits organized among RCDs, local governments, and other agencies) are also beginning to bank potential carbon credits through tree-planting programs they underwrite. In Latin America, The Nature Conservancy has used sales of forest-based carbon reduction credits to fund additional research on tropical forests and contribute to the cost of acquiring certain reserves. The government of Costa Rica is using the sale of forest-based carbon reduction credits to acquire clear title to public forest reserves, as well as to acquire CEs on key private forests.

Forest Watershed Services

As discussed in Part One, water supplies for drinking, agriculture, and energy generation often flow from upland forested catchments. The ability of forests to catch and filter water and moderate flows is of great value to society and our economy. Forest owners and managers have water quality mandates to fulfill under the Clean Water Act. However, owners committed to high levels of protection within municipal, hydroelectric, or other key watersheds are gaining access to a new market for their stewardship through the sale of CEs to water districts, utilities, and similar public agencies. In this way private owners can be rewarded for the quality of their exported water.

Most municipal watersheds that are protected are in public ownership. In the last twenty years, CEs have begun to be used as an alternative to fee title acquisition to protect privately owned watershed lands from incompatible development and intensive road building as well as guide forest management to maintain tree cover and reduce erosion or mass wasting risks. The growth of local funding programs, such as those described below, will expand the market for forest watershed services and accelerate important forest conservation.

Payment for the acquisition of title or interests in watershed forestlands has generally come from either general tax revenues, water or electricity fees, or sometimes a surcharge or excise tax. (Fees or taxes for users also encourage water conservation.) When the federal Safe Drinking Water Act was amended in 1996, each state gained access to federal funds for improvements to public water supply systems. Under the act, states can designate up to 10% of their federal grants for loans to acquire land or CEs in watersheds to prevent contamination of supplies. The loans are then paid back through user fees or similar mechanisms.

The St. Johns River Water Management District in northern Florida is actively acquiring CEs on private forests in its watershed, including a 36,000-acre easement granted by Nekoosa Packaging Corporation. In Spokane, Washington, residents pay a surcharge of $15 per year to the city for acquifer protection. The city of Providence, Rhode Island, receives $1.29 per 100 gallons of water used from a state surcharge. The city uses 55% of these funds to buy watershed lands from private owners. In so doing, it has increased the protected acreage tenfold. New York City recently initiated a program to upgrade its municipal water quality through improved watershed management at about 20% of the cost of building a filtration facility. The program resulted in users' water bills rising by only 9%, rather than the predicted doubling. Communities across the Unites States have yielded similar savings.

In the western United States where water is always scarce and water rights are well established, a new water market is developing where holders of rights (often of water flowing from public lands through publicly funded water aqueducts) are selling or leasing those rights to downstream users, usually metropolitan water districts needing to supplement their own rights. Local governments in the mountains, upstream forest owners, and agricultural enterprises are beginning to sell water rights to generate revenue. Some conservation groups, such as Water Watch and the Oregon Water Trust, both in Oregon, are buying water rights to dedicate instream flows to habitat. Such transactions can provide an incentive for conservation of watershed lands and improved land use practices to protect water quality. Dedication of water to instream flows can be permanently secured through a CE.

Public and private electric utilities have a major economic interest in protecting watershed lands as well. Watersheds that are managed for water quality and sedimentation level can thereby avoid their own costly filtration systems, extending the life of their reservoirs. Further, in well-forested watersheds peak flow is reduced and run-off time extended reducing the threat of spillover. One of the best examples of the potential market among power users for watershed protection is from South Asia. Most of Bhutan is a steep, forested watershed. Hydroelectricity is one of the country's major exports and neighboring India, with its growing demand, is its primary customer. Protecting Bhutan's forest cover is essential to maintaining the stability of its soils and the functioning of its power reservoirs. By selling hydroelectricity at a price of $.07/kw and producing it at a cost of $.0237/kw, Bhutan is able to reinvest in watershed protection and extend the life of its installed power capacity. Forest stewardship also helps maintain the country's significant ecotourism trade.

Expanding Capital and Markets for Forest Conservation

Greater investments of public and philanthropic capital are needed to fund the protection of private forests for their intrinsic, noncommodity value to society. Conservation of biodiversity and wildlife habitat values in private forests has no ready financial markets. Public and philanthropic entities are the natural sources for such investments because, unlike private capital, they do not demand a competitive financial return. In this public marketplace for private forest conservation a rational pricing system generally prevails. That is, the longer the period of commitment to conservation, the greater the price paid for that commitment (or "the more you give, the more you get.")

Increased public demand for forest protection fuels greater public investment in the conservation value of private forests. This in turn can leverage greater private investment in forest stewardship by lowering the cost of capital for investors and bringing in needed cash for many family ownerships. With the trend of increasing forest disposition by industrial and nonindustrial owners alike, there is a historic opportunity—if not a necessity—for expanding the availability of capital for forest conservation to influence the results of these transfers.

The good news is that as the threats of sprawl, fragmentation, and intensive development have grown throughout the country, funding for land and resource conservation has been increasing from local, state, and federal sources. Communities are realizing that the costs of development literally exceed their returns. "Land conservation is often less expensive for local government than suburban-style development," notes urban planner Holly L. Thomas in the Trust for Public Land's 1999 report, *The Economic Benefits of Parks and Open Space.* "Farms and other types of open land, far from being a drain on local taxes, actually subsidize local government by generating far more in property taxes than they demand in services." According to a report prepared by the Land Trust Alliance, 84% of the 184 open-space initiatives on the ballot in 1998 passed (Gilmore et al. 1999). States such as California, Florida, and New Jersey are dedicating billions of dollars to protection of remaining open lands, increasingly funding CEs acquired by agencies and land trusts. Public funding mechanisms for conservation include issuing general obligation bonds, creating or increasing real estate transfer taxes, and instituting sales tax surcharges. Governments are also increasing direct budget appropriations, especially at the state level.

The Sonoma County (California) Open Space and Agricultural Protection District, created by referendum in 1990, is funded by a sales tax surcharge of one-quarter of a percent that is generating more than $10 million annually, with the twenty-year stream of funds dedicated to land conservation in this fast-developing county north of San Francisco. As of spring 1999, the district had acquired CEs on more than 30,000 acres of land at a cost of about $38 million (Fricker 1999). Some communities, such as Crested Butte, Colorado, are so committed to conservation that more than one hundred merchants are collecting a 1% voluntary sales tax and donating it to local nonprofit land conservation organizations, raising $100,000 in 1998 alone (Lerner and Poole 1999).

The bad news is that much of this funding is in areas already experiencing tremendous development, thereby increasing the costs of open-space protection. Those who want to expand private forest conservation can learn from the example of Boulder, Colorado, which in 1967 became the first city to pass a dedicated sales tax to fund the preservation of open space—well before Boulder became a modern boomtown. Thanks to this farsighted action, Boulder today enjoys 40,000 acres of open space, much of it in greenbelts surrounding the city. Many credit this public investment with the attractiveness—and high cost—of Boulder real estate. Boulder's contained development, drawn by the amenity value of protected lands, generates property taxes that have paid the costs of open-space acquisition many times over (Lerner and Poole 1999).

The plight of private forests is not yet apparent to many. As the benefits of Forest Legacy and state-level CE funding programs for private forestlands begin to be appreciated, however, funding is increasing. The challenge is raising the visibility of forests to increase public investment—and using the funds to get ahead of the wave of development, leveraging more conservation of private forests with multiple funding strategies before the bulldozer is next door.

In addition to public agencies, nonprofit land trusts are the leading direct conservation service providers. In fact, many private forest owners, especially in the West, would prefer to work with nongovernmental entities as long-term partners. As discussed earlier, land trusts are increasingly cooperating with forest owners' associations and others to meet the needs of private owners. Land trusts are also in a position to flexibly mix and match funding sources to leverage the greatest conservation gains. A number of states, including California and Vermont, rely primarily on land trusts for the implementation of their Forest Legacy programs.

The public market for permanent private forest conservation can be enhanced in a number of ways. Concerned citizens need to better understand and utilize state and local funding mechanisms. Increasing appropriations to Forest Legacy and other federal conservation programs need to be supported. Nonprofit land trusts need to expand their access to public capital for the acquisition and management of forest CEs. And new public funding mechanisms need to be developed, whether through innovative local transfer taxes, development rights banking systems, or state and local conservation tax credits. One new capital source that is being promoted uses the tax-exempt bond market. Drawing on the model of nonprofit hospitals, with a change in the Internal Revenue Code, nonprofit forest conservation organizations could be established that issue tax-exempt bonds to capitalize the acquisition of forestland at low interest rates. Under proposed legislation, acquired properties would be subjected to CEs to ensure that this public subsidy is not abused and that the public interest in ecosystem protection is furthered. Nonprofit owners would then repay the bonds from revenue generated through stewardship management of the forest properties.

As discussed further in the final chapter, although growing public funding is essential to fueling a conservation marketplace, philanthropies can play a major role as well. Philanthropies and other charitable donors can expand the conservation marketplace by directing more funds to CE acquisitions individually or through pooled grant-making funds. Some innovative foundations are making grants or low-cost, long-term loans (called program-related investments) to land trusts for *conservation easement project revolving funds*. Using these revolving funds, land trusts can act more quickly in the real estate marketplace and secure the protection of important forest properties. Then the funds can be paid back in time through a variety of means, including allocation of public funds, dedication of a portion of the conserved property's sustainable timber harvest receipts, or resale of the property to a private conservation buyer. The New England Forestry Foundation, for instance, has been able to expand its forest conservation by having access to a $1 million credit line financed by a foundation. At an interest rate of 0 to 4%, it can utilize the funds for up to five years, with repayment of principal and interest at the end of the term. PFT is increasing funding for its Strategic Opportunities Conservation Fund from $5 million to $10 million so it has sufficient working capital to partner with private and public entities to leverage larger-scale conservation.

Expanding Fiscal Incentives for Conservation

As described in chapter 3, aspects of tax policy can be an impediment to long-term forest stewardship. Although it is beyond the scope of this book to make detailed recommendations about specific changes to tax policy, there are many potential fiscal incentives that center on enhancing returns from forest conservation and eliminating tax barriers to stewardship investments. We would like to highlight several of the numerous proposals under discussion to give our readers an idea of their range.

Allow landowners to more fully realize the income tax benefits of conservation easement gifts. Conservation easements on forestlands are often of significant monetary value, and landowner donors may not have sizable incomes against which to offset the value of this gift. The Land Trust Alliance and others have proposed increasing the realized value of the CE gift to the donor by increasing the amount of the deduction that can be taken for the easement donation from the current limitation of 30% of adjusted gross income to 50%. They have also proposed eliminating the five-year limitation on donors' ability to carry forward the unused portion of the deduction.

Create tax credits for conservation and stewardship. Several states, such as California, Colorado, and North Carolina, have recently enacted state-level tax credits to help landowners receive more cash value for their CE donations. Although programs vary from state to state, in general the tax credit is allowed for some percentage of the value of the CE and can be taken in addition to the existing charitable tax deduction. Tax credits for landowners' out-of-pocket costs for creating CEs would also be beneficial. Other proposals in circulation would provide some level of dollar-for-dollar tax credit for forest owner investments in certain key forest stewardship activities, such as restoration of highly threatened forest types, enhancement of habitat for threatened and endangered species, and removal of poorly placed and highly erosive or unstable roads and log decks.

Put conservation sales of property and conservation easements on a more competitive footing with sales for development. Often conservation organizations cannot pay full fair market value for the conservation acquisition of a particular property or CE. Even when they can, landowners often need an additional incentive to make that commitment to conservation rather than further development. A proposal to exclude from capital gain taxation 50% of the income from the conservation sale to land trusts or government agencies was introduced in Congress in 1999 and gained considerable support.

Expand the existing estate tax benefits of creating conservation easements.
Under Section 2031(c) of the Internal Revenue Code, properties within a certain range of metropolitan areas, national parks, urban parks, and wilderness areas—which is to say, near developing areas—can receive special estate tax treatment. This section exempts from estate taxes the value of underlying land (exclusive of retained development rights) of a taxpayer who has established CEs on his or her property up to certain limits, if it qualifies under the geographical requirements. The LTA has proposed that Congress expand this incentive to all areas of the country and eliminate the cap on the value of the exemption.

Reduce negative impacts of estate taxes on forestlands. A small but significant portion of forest ownerships is at risk of excessive harvest or fragmentation owing to the effects of unfunded estate taxes. Although there is great debate over the extent of the threat, it clearly looms over some larger, well-stocked forest ownerships. Proposals are pending that include increasing the existing estate tax exemption to up to $5 million so that the tax burden falls only on the wealthiest ownerships (those most equipped to either fund the tax or create a high-quality estate tax plan); and/or accelerating the phase-in of the exemption increases already in place. There are also proposals to simplify the requirements for the special-use valuation, which reduces the appraised value of farms and forests under certain circumstances and allows low-interest installment payments on taxes due from any taxpayer.

Reduce the impact of capital gains taxes on long-term forest investments.
The longer a forest owner holds a naturally appreciating asset such as timber, which grows every year regardless of the market, the more likely that capital gains taxes will bite disproportionately into the gain when it is finally time to harvest. That is because of the compounded impact of inflation over the many decades a family may hold its timber. If the timber basis is indexed to inflation after twenty years, taxes will not eat up the rewards of long-term stewardship for landowners committed to restoring and managing older age classes of trees.

Improve cost recovery for forest stewardship investments by smaller forest own-ers. In chapter 3 we reviewed the *passive investor* tax trap that prevents many NIPFs from fully deducting management and property tax expenses as they occur. A change in the definition of *material participation* that recog-nizes more periodic investments of time by forest owners would effectively eliminate this hurdle and encourage greater stewardship investment. Such a change could be accomplished by lowering the annual number of hours required for forest management and/or averaging hours over a period of time, or by other means.

Provide for the tax deductibility of noncommercial forest stewardship expenses. Perversely, many stewardship expenses are not considered by the IRS to occur in the normal course of business because they are not incurred with a profit motive. Such expenses include voluntary ecosystem restoration activities encouraged, for instance, through the federal FSP; for instance, use of prescribed burning, control of invasive vegetation, restoration of riparian vegetation, streambank stabilization, road decommissioning, and many other investments in habitat or water qual-ity enhancement. Several proposals for changes to the Internal Revenue Code would allow forest owners to deduct such expenses against cur-rent income. In addition, these changes would allow owners who invest in reforestation without the intent of timber harvest to claim the exist-ing reforestation tax credit and amortize the cost of the trees against ordinary income.

Improve property tax treatment for forestland. Property tax policies of cer-tain states and localities that are contributing to the fragmentation and conversion of forestlands need to be addressed. In the survey conducted for this book, fears of property tax increases were voiced more frequently than concerns about estate taxes. For instance, the ad valorem tax on standing timber needs to be eliminated in the few states where it remains. Ad valorem means that timber is taxed at its standing value, whether it is cut or not. Having to pay tax on an illiquid asset is a real incentive to liq-uidate that asset, so many states have moved to some form of preferen-tial tax treatment for forestland, as described in chapter 4. However, even states that have moved to the newer tax system need to institute more read-ily accessible forest-use tax programs if they are serious about maintain-ing the base of private forestland in their jurisdictions.

Improving Forest Liquidity Options

As discussed in chapter 3 lack of working capital for stewardship is a significant barrier for smaller and medium-sized nonindustrial ownerships. These owners often lack sufficient liquidity to invest in property stewardship—even in the costs of establishing a conservation easement. New mechanisms and expanded programs are needed to meet up-front stewardship costs and provide for cash flow between timber harvests. This can be accomplished in a number of ways, as discussed below.

Funding government cost-share and technical assistance stewardship programs such as SIP, EQIP, and Partners for Fish and Wildlife at levels commensurate with landowner demand is one obvious and attainable approach.

The Oregon Forest Resource Trust model is an approach to learn from and expand. Through it public funds are loaned to NIPF owners for reforestation of historically understocked forestlands based on landowners' commitments to repay the funds when the lands are ultimately harvested. Similar state-level forest stewardship loan funds could provide advances on future timber harvests for a variety of habitat restoration projects and other activities providing public benefit. Such funds could draw on both public and charitable sources.

A timber futures market could be created through either a public lending program or a government-guaranteed bank loan program, or by development of a private-sector commodity product that could be sold on exchanges such as the Chicago Board of Trade. Such a mechanism could allow the landowner to obtain the future value of a timber harvest, discounted at a rate to account for the cost of money (similar to prepaid interest) and various risk factors, such as possible market downturns, natural disasters, or future regulatory constraints. If such a futures market were more closely tied to a conservation outcome, using the public program approach a seller would be allowed a lower discount rate if the property were conserved and sustainably managed.

Small landowner financing innovations such as The Nature Conservancy's pilot Forest Bank program are worthy of support and emulation. The Forest Bank was recently introduced in Virginia's mountainous Clinch Valley, where the average forest ownership is 80 to 120 acres in size. Although the structure is still evolving, in general this novel mechanism allows forest owners to contribute their timber rights to the bank (characterized as "depositing" them) in return for a guaranteed annual dividend based on the initial value of those rights, likened to a return on a bank certificate of deposit. Members can withdraw their contributions at a reduced value. The Forest Bank creates a sustainable (and certified) forest management plan for the property, harvesting and marketing timber through time. The collective pool of timber rights exercised in this way is meant to fund the guaranteed distributions to participating landowners. In this way landowners trade in their periodic timber receipts—if they were managing their land for timber at all—for a more consistent flow of revenue. The bank pays all management costs and makes all stewardship investments. This addresses both small landowner illiquidity and lack of knowledge or means to invest in stewardship forestry. Similar to a CE, contribution of timber rights to the bank ensures that the land will remain forestland and be well managed.

Additional public benefits possible from the Forest Bank include providing sustainable supplies of timber for growing value-added enterprises in the region; ensuring that these forests, many of which are on unstable slopes in a high-erosion area, are managed for improved water quality; and providing a critical mass of forest properties that can be managed for long-term values as a buffer to The Nature Conservancy's protected areas in the watershed. Therefore, like other forms of cooperative ownership, the Forest Bank could be a good tool for landscape-level forest protection in certain areas.

Lending programs that finance costs of stewardship and conservation, such as for forest stand improvement, habitat restoration, and establishment of CEs, can also be instituted by community organizations. For instance, such programs can be created by community development corporations, community revolving loan funds, nonprofit sustainable development entities such as Shorebank Pacific Enterprises or MACED, and even commercial banks such as Shorebank Pacific itself ("the first environmental bank," focused on conservation-based development in the coastal temperate rain forest). It may be possible for a forest community to persuade its local bank to fulfill some of its requirements under the Community Reinvestment Act through the creation of a stewardship lending program.

Creating New Stewardship Ownership Forms

As discussed earlier, many forms of forest ownership are not well aligned with the biological time frames of forests and the complexity of their natural assets. Both legal form and frequency of turnover in ownership can profoundly shape a forest. New stewardship-oriented forest ownerships could significantly contribute to reversing the cycle of forest degradation, fragmentation, and loss. The advantage of experimenting with new ownership forms is that successful ones could attract significant new inflows of private capital into forest conservation. With millions of acres of forestland now in play, it is important for conservation organizations and private nonprofit conservation partnerships to build their capacity to intervene in the marketplace to generate conservation outcomes. New ownership forms, including those that combine nonprofit and for-profit entities, could play significant roles. The forms we will highlight below utilize CEs as mechanisms to ensure continuity of forest stewardship and prevention of property conversion through time. These examples of different approaches can be hybridized, and represent a selection of emerging ideas.

Forest Investment Management Organizations

At the larger scale of ownership, a variation on institutional management through TIMOs is possible. PFT has proposed the creation of FIMOs, or *forest* investment management organizations, with similar tax and investment advantages. An FIMO can organize funds of investors, or manage dedicated accounts for larger institutions, to provide competitive risk-adjusted returns through the acquisition, conservation, and management of forestland on the stewardship forestry business model. FIMOs would seek to capitalize on the suite of ecological goods and services available from its portfolio of forest properties. In the acquisition process, the FIMO would ensure that CEs would be established on the core forest properties to ensure that they are not fragmented or developed in the future, and that future forest management would enhance and restore forest ecosystem wealth.

FIMOs can organize capital from a variety of sources, including private individuals and institutions as well as public and philanthropic entities. Well-capitalized FIMOs would be in a position to participate in the fast-changing forestland marketplace, seeking to create more conservation outcomes as properties change hands. While FIMOs are likely to sell portfolio properties at some future date, CEs would protect forest asset values and provide for management consistency under the new owners. Ownership units or shares in an invesment portfolio managed by an FIMO might at some point be organized for marketing to the public, providing a more liquid market for units while protecting underlying forest assets.

PFT has developed a business plan for a prototype FIMO, Cascadia Forest Stewardship Investments, to operate in the Pacific Northwest.

Forest Stewardship Communities

Although still hypothetical, this ownership form for community forests has promise for conserving some medium- to large-sized forest properties while providing for limited residential development and sustainable forest management. A forest stewardship community would consist of a residential cluster joined with an adjacent forest property. The forest tract would be owned by the members of the residential community and secured from fragmentation and development by a CE. Here "the community" refers to an association of residential landowners, each owner vested with an undivided interest in the larger forest property. Individually and collectively the residential owners could enjoy and utilize the forest, consistent with the CE terms as well as the community's governing covenants and restrictions. As with a condominium development, a board would be charged with managing the properties and ensuring forest stewardship. Community members would pay annual fees and receive their share of revenues if any. Their community interest would be marketable, comprising joined residential and forest rights, subject to the restrictions that would travel with the property. Such forest stewardship communities could be beneficially created in forested rural areas that are experiencing growing residential demands. Through this ownership form, non-forest development and infrastructure would be clustered in the most efficient and environmentally sensible fashion, while the major forested tract would be conserved and managed for its long-term ecological values. Sale of the community membership units would fund the long-term conservation of the forest.

While forest stewardship communities could be organized by for-profit, conservation-minded developers, they could also be developed by nonprofits or innovative local agencies, such as resource conservation and development councils (RC&Ds). The latter two types could be financed conventionally, or possibly through the issuance of tax-exempt development bonds to acquire the land, design its conservation-based development, and resell units to private residential buyers as above, enabling repayment of the bonds. There are many potential variations, some of which could entail public subsidies to ensure investment by low-income local residents.

Forestland Cooperatives

The forest cooperatives discussed in chapter 5 could evolve into actual mergers of interest among like-minded landowners. This kind of cooperative ownership would be a means to reassemble a forest landscape already fractured into multiple, relatively small ownerships. As with the other forms described here, it could also establish continuity in stewardship across a larger scale than would otherwise be possible. In addition, merging ownership interests through formation of a cooperative could help address landscape-level impacts of an individual property's timber harvest and other land uses; even out cash flows for individual owners sharing in the collective revenue stream; provide economies of scale in management and product merchandising not readily available to individual owners; and engage individual forest owners in stewardship practices they might otherwise be unaware of or unable to invest in. Legally such an ownership could take the form of a cooperative, with ownership shares, or of an partnership or LLC. Landowners would contribute their entire ownership interests to the cooperative, or simply the fee titles, while retaining rights to their homes. The latter could be sold separately, again as with condominiums. Short of legally merging the ownership interests, the cooperative could be structured as a federation or affiliation of owners who make a binding agreement to common management, but retain their separate ownerships. Either would be best focused on a common watershed. Conservation easements could be used to embed the accepted forest stewardship on the ground and in the deed, so that future owners would be bound by the same goals and restrictions on incompatible uses. In effect, such a cooperative would be a forest stewardship community created from a group of existing forest ownerships.

Nonprofit Forest Conservation Ownerships

As described in chapter 1, nonprofit conservation organizations have typically acquired fee title on forestland either as acquisition agents for a government entity or for protecting an ecologically significant and well-functioning forest property as a quasi-public preserve. A handful of conservation groups—the Society for Protection of New Hampshire Forests and the New England Forest Foundation, for instance—have over the decades accumulated tens of thousands of acres of forest under their management. Over the last few years a number of nonprofit organizations have begun to purchase or otherwise acquire managed forest properties. They conserve the properties while maintaining timber harvest and other productive uses compatible with protection of ecological assets, such as threatened wildlife habitat.

In the Northeast about 500,000 acres of private timberland have been acquired in several transactions by the Vermont Land Trust, The Nature Conservancy, and the Conservation Fund. Portions of these properties are being held and managed as demonstration forests for sustainable forestry and conservation purposes. Owners who retain land trusts are creating a new form that combines high standards of forest ecosystem protection with commercial forestry. In so doing they hope to generate surplus revenues over the stewardship costs of forest property for protection of more forestlands. Further, they hope to create data and working examples of stewardship forestry as models for other private forest owners interested in restoring ecosystem wealth. To expand this concept, however, increased philanthropic working capital is needed as an anchor for banking and other financing arrangements. (Further description of their transactions can be found in appendix D.)

Nonprofit corporations, like their for-profit counterparts, do not have limited life spans. Because they are also dedicated to the public benefit, they could form a good fit for the long-term management of forest ecosystems. However, even nonprofits require good governance policies and procedures to ensure that the forests under their ownership are being managed for ecological asset values and not simply generating revenues to be used for other charitable purposes. Conservation organizations have been criticized for using charitable means to acquire properties that they then resell, unencumbered, for fair market value to fund other projects. Therefore, even under nonprofit ownership, establishing a CE to restrict nonforest development and guide long-term stewardship is a good practice.

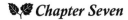 *Chapter Seven*

An Action Plan to Accelerate the Conservation of Private Forests

The data and trends clearly indicate the increasing level of threat to the existence of the seemingly vast landscape of private forests. As we have seen, three primary forces combine against them: degradation, fragmentation, and conversion. While the complexity of interactions between these forces can be daunting, many programs are available to address them. Still, sustaining a strong, organized counterforce of conservation is a formidable challenge. In this chapter we outline a strategy for concerned forest stakeholders to mobilize and focus scarce resources with the goal of accelerating and expanding the conservation of private forests to forestall further forest loss. Accomplishing this goal will be possible only if all concerned, from forest owners to the general urban public, contribute. We believe that with a cooperative effort at local, state, regional, and national scales, America's private forests can thrive in the new century.

Establishing Conservation Objectives for America's Private Forests

Such a cooperative plan of action needs to focus on a few major objectives that address the fundamental threats to private forests. These objectives can provide a compass by which forest stakeholders can guide their actions and make choices about how to invest their resources.

Turn the Tide on Private Forest Loss. In many key forest areas loss is accelerating with burgeoning populations and sprawling development. Therefore, key threatened forest areas need to be identified and their protection expanded.

Dramatically Reduce the Fragmentation of Larger Forests. Although forests are caught up in a huge current of encroaching development, the process of fragmentation—which degrades and diminishes forest functions—must be slowed. Therefore we need to focus conservation efforts on larger, intact forest properties.

Create Ways to Functionally "Reassemble" the Landscape. With more than half of U.S. private forests already in ownerships of fewer than 500 acres, landowners need to be encouraged to cooperate in forest stewardship across ownership boundaries so that the pieces of the landscape can begin to be put back together again.

Fuel the Restoration of Ecosystem Wealth. The decline in a variety of measures of forest health—timber characteristics, presence of threatened and endangered species, level of biological diversity, and water quality status—all indicate the tremendous need for public and private investment in private forest ecosystems to enable them to provide the goods and services our society needs and desires. Consequently forest landowners' and communities' access to stewardship capital and resources needs to be expanded.

Build a Culture That Values Forests. Reconnecting the general public, and urban interests in particular, to the values and contributions of private forests is essential to the long-term conservation of forest ecosystems in the United States. Therefore concerned forest stakeholders need to work together to make private forests more meaningful and relevant to the everyday lives of more Americans.

As described further below, although the current intensification of turnover in forestland and shifts in ownership may threaten some forests, it also offers an unprecedented opportunity to accomplish conservation. New alliances of public, private, and nonprofit partners can participate in the changing marketplace and collaborate on larger-scale conservation projects. It is not enough, however, for the scale and scope of conservation projects to increase in the short term. A new cultural basis for ongoing forest conservation needs to be built so that, in the longer term, many of the threats apparent today are reduced or eliminated. Accomplishing the objectives described here will require significant, ongoing investment of time, money, and collective will on the part of everyone concerned about forests. Therefore, it is important to remember that the negative trends affecting forests and diminishing forest biodiversity are complex and have accumulated over a long time. Solutions are correspondingly multifaceted, will take time, and will evolve—even while many positive changes can be achieved quickly.

We acknowledge that certain threats are not directly addressed in the strategy discussed in this chapter. Some, such as the impacts of invasive exotic species, population growth per se, and the need to create more vital and attractive urban centers, are of profound concern. Clearly, efforts in these areas also contribute to effective solutions.

Everyone Can Make a Difference

In the effort to preserve America's private forests, everyone has a role to play. Active engagement of the people who own and manage private forestlands is essential to successfully implement this strategy. Forest conservation needs to make sense to them and help them fulfill their goals of forest ownership. Therefore, the approach described in this chapter focuses on methods and tools that will help forest owners be stronger engines of conservation. As there are different kinds of owners, with differing goals and management contexts, each requires appropriate tools, although some solutions may be shared. In general there are three kinds of forest owners: institutional nonindustrial owners and industrial owners, who usually have the largest ownerships and who own forests as financial assets; mid- to large-scale nonindustrial owners who hold forests primarily as financial assets but also as personal assets; and small NIPFs who typically own for personal reasons, and do not generally manage forests for financial return on a regular basis.

Additionally, forest owners' key influencers are important participants. These are parties whose actions or thinking play a critical role in determining or constraining landowner behavior. They include advisors and service providers such as foresters and loggers, attorneys and financial planners, and real estate agents. Their inertia or creativity can substantially increase or hamper conservation efforts. On the other hand, their involvement can grow as they understand the financial benefits of conservation-based forestry.

Policymakers and public officials at the local, county, state, and federal levels also play important roles in influencing the future of private forestlands. In fulfilling their policymaking and regulatory roles, they can either promote or discourage private forest conservation and stewardship. They are also critical to providing the regulatory framework for new markets, and for determining public expenditures.

Last but not least, the general public is essential to ensuring that private forests continue to provide their many benefits. Their perceptions of the contributions that private forests make and of the quality of private forest management are critical to enabling increased forest conservation at all levels. Without public support, forest landowners face an uphill battle in forest management. Without public support, policymakers and public officials will not fight for appropriate land use policies, tax reform, or increased funding to advance forest conservation goals.

Funding the Expansion of Private Forests Conservation

Conserving forests and restoring ecosystems can cost substantial amounts of money. To successfully expand forest conservation, funding also needs to expand. Three basic kinds of funding are available for investment in the restoration and conservation of private forests, each with its own motivations and requirements. Public funds—derived from tax receipts or bond measures and administered by government agencies—provide a substantial source available to protect the public benefits of private forests. Another much smaller but public-spirited pool of capital is that of philanthropies, the charitable gifts of foundations or individuals. Finally, there is the huge pool of private capital that can be provided by forest owners and investors. Public and philanthropic capital is invested in private forests to secure public benefits and does not seek financial returns. Private capital, however, requires financial returns, which it can gain while also rebuilding ecosystem wealth. Each source of conservation funding has an important role to play in expanding the conservation of private forests and needs to be augmented as part of any conservation plan.

Focus on Private Forests at Risk

Based on our analysis of the data, certain kinds of forests should form the focus of cooperative conservation efforts because of their higher degree of risk. They are discussed below.

Larger Forests Everywhere

Regardless of location, any of the estimated 68,000 ownerships of 500 acres or more, representing 45% of the forest landscape, are most at risk of fragmentation into smaller properties. This is evidenced by the parcelization of 22% of their area into properties of 100 acres and less between 1978 and 1994 (with the bulk of the loss in those ownerships between 500 and 1,000 acres). These ownerships are most likely to include our most intact forests, with larger contiguous parcels. By facilitating the conservation of these larger ownerships we can get upstream of the problem to its source and leverage the greatest conservation.

In particular, larger forest ownerships just beyond the direct influence of growing metropolitan areas can make an excellent strategic investment of conservation resources. These properties are at great risk because they are next in line for development. This is especially important in the vicinity of medium-sized cities that are the coming generation of major metropolitan areas. Yet because these forests are still beyond the urban-influence zone, their conservation can be accomplished at less cost today than tomorrow, providing the greatest financial leverage. Real progress in forestalling forest loss ten and twenty years from now will be made by focusing on areas where forest values are still strong enough to outweigh real estate development values, and where fragmentation has not already overwhelmed the landscape.

Also at high risk are larger forest tracts that are in ownership transition, whether through corporate restructuring or family succession. Larger forest tracts that are near public lands or along the water are also highly vulnerable to conversion to rural residential and recreational parcels.

Larger forests with well-stocked, older natural stands are perhaps doubly at risk because they are both increasingly rare (and highly valuable for biodiversity) and increasingly valuable to timber companies to provide near-term supplies. Interestingly, larger, poorly stocked forest tracts are also at high risk because they are candidates for near-term fragmentation if owners cannot afford the investment of time and cash to restore their timber and ecological values.

Other at-Risk Forests

In addition to ensuring that larger forests stay intact, we must protect medium-sized and strategically located smaller forest properties. Forest ownerships of between 100 and 500 acres are at particular risk of conversion into ever smaller parcels, as evidenced by the 18% of area lost from this size class between 1978 and 1994. They are also at risk of greater ecological degradation due to many owners' lack of motivation and resources to invest in forest stewardship.

Rare larger and medium-sized tracts in key metropolitan areas, such as Atlanta, Seattle, and Washington, DC, have important conservation value because of their scarcity. Their signficant public benefits can warrant the greater expense of conservation, as compared with more distant properties. In fact some forest ownerships of 50 acres or more that provide important ecological values within their local or regional landscape should be identified and conserved. Finally, forest ownerships of between 50 and 100 acres typically need more active stewardship to restore ecological and economic forest values. Therefore, they are an important focus of conservation investment to reduce the risk of further fragmentation and outright conversion.

Using the Conservation Toolbox

Before we outline a plan of action to better accomplish the conservation objectives put forward here, let us review briefly the nature and uses of the conservation tools described in chapters 5 and 6. Forest degradation, fragmentation, and loss operate on different time scales and different tools and resources are required to address them. Generally, the more imminent the threatened loss of forestland and the more advanced the fragmentation, the more financially driven and immediate the solutions will need to be. Therefore, in these circumstances the tools to apply are those described under *Markets and Financial Mechanisms* in chapter 6. These tools are largely targeted at owners who manage their forests as financial assets, and secondarily at policy and decision makers. Using financially based tools such as these, significant near-term progress can be made in conservation of specific forestlands, and new conservation markets can be catalyzed that will fuel future protection.

On the other hand, where the primary concern is ecological degradation but the forest area is relatively intact, more process-oriented solutions are needed, as described in chapter 5. To secure lasting, large-scale conservation of the forest landscape, the human culture that determines the fate of that landscape must be addressed. Cultural tools are used primarily to influence the stewardship commitment of owners who manage their forests as personal assets. They also influence the public, advisors, and service providers. Public policy decision makers are a secondary audience. These tools and initiatives are also useful to new partnerships and coalitions that are working on community- and landscape-level initiatives.

The condition of forests, the nature of forest ownership, and the relative severity of forest threats vary from region, to region, from state to state, and from watershed to watershed. Therefore, the broad approach articulated here can be effectively supplemented by more detailed strategic planning processes at regional and local scales that should involve the range of forest stakeholders. What we are presenting can operate at these different scales of time and space, adapted to local contexts. In each region, problems will be perceived at varying levels of importance to differing audiences. Thus at the local level, preventing the loss of a specific forest may be more important than restoring degraded forests, and fragmentation as a landscape pattern may not be understood as a problem by all concerned. Immediate problems, such as threatened development for a specific forest, call for immediate actions. Other problems are midterm—occurring within 5 to 10 years—such as increasing impacts of parcelization from unplanned breakup of forests or lack of zoning. Such events call for more process-oriented solutions. There are also long-term problems that may not reach a crisis level for decades, such as declines in biodiversity or increases in sedimentation. These problems require cultural change as expressed in new management paradigms, in addition to long-term, on-the-ground restoration efforts.

A Strategic Conservation Plan

Given the objectives we have outlined, we propose that the conservation of private U.S. forests can dramatically expand if concerned forest stakeholders cooperatively implement the strategic actions described below. Throughout this discussion we will be referring to tools and approaches described in chapters 5 and 6, which are a useful cross-reference. Table 7-1 provides a synopsis of the main activities that we believe will accelerate the conservation of private forests. This plan has two major thrusts: bringing the conservation market to scale and integrating conservation into forestry, and forestry into society. The latter emphasizes markets and financial mechanisms while the former emphasizes cultural change tools.

Bring the Conservation Market to Scale for Rapid Gains in Private Forest Protection
While there is an existing conservation market, it is often inefficient and insufficient on larger scales. Increasing financial returns of conservation and stewardship is critical to compete with the financial returns of

development and degradation. Expanded conservation of private forests requires the creation of a greater capital market for direct conservation through increased philanthropic, public, and private funding, as well as development of new markets for ecosystem services. The counterpart incentive of increased conservation revenue is reduction in the cost of capital for owners to undertake conservation-based management. Therefore, in addition to developing and expanding the market for forest conservation, key changes to fiscal policy are needed. The following specific actions should be undertaken.

Provide new conservation capital to intervene in dispositions of significant forest properties. As a society we are at a historic moment where dispositions of IPF and NIPF forestlands and conversion pressures are intensifying. Although these trends increasingly threaten forests, they also provide tremendous opportunities for new partnerships and market interventions that can yield immediate and direct conservation. As described in Part Two, The Conservation Toolbox, under Expanding Capital and Markets for Forest Conservation, increases in philanthropic support can leverage greater private and public investments to fund the conservation value of forestland transactions. Such events will allow private-sector "conservation buyers"—whether forest-products companies, pension funds, or partnerships—to partner with philanthropies, nonprofits, and the public sector to permanently protect larger-scale forest properties.

Increased public/charitable investment will enable larger financially driven forest owners to reduce their cost of capital by having CEs purchased on their lands (whether at fair market value or at a bargain price).[1] This could make a difference in the ability of some owners, such as pulp and paper companies, to retain more extensive acreage in the United

---

1. Although perpetual CEs are not the only possible means to provide some value to landowners for their forests' conservation value, they are the most widely accepted and legally established means. Making a permanent commitment to conservation maximizes the return to both landowners and the public. Term-limited forest conservation rentals, similar in concept to rentals under the Conservation Reserve Program, would provide less income to landowners and correspondingly less gain for the public. Therefore, we believe scarce dollars should be directed only to expanding the use of CEs on managed forestlands to provide the best return on investment.

Table 7-1.

Matrix of strategic actions to accelerate private forest conservation

Actions	Focus Audience
Bring the Conservation Market to Scale	
Expand public and philanthropic capital for conservation easement acquisition costs; create regional funding pools	Larger forest owners and properties
Provide low-cost working capital from public and philanthropic sources to conservation organizations to participate in the forest disposition marketplace	Larger forest owners and properties
Expand public funding for forest conservation easement acquisition and forest stewardship programs	Policymakers, public, all forest owners (local, state, federal)
Develop and invest in prototype forest carbon banking mechanisms	Policymakers, all forest owners, carbon credit buyers
Expand water markets to provide returns to forest owners for conservation and stewardship	Water districts, policymakers, forest and "downstream" communities, forest owners
Support changes in tax policy to improve returns from conservation and stewardship	Policymakers, forest owners, public
Invest in new public and private financial mechanisms to provide improved liquidity to forest owners	Small- and medium-sized forest owners
Expand market-making, R&D, educational efforts to increase returns for biodiversity-oriented forest management	Buying public, forest owners, processors, researchers, educators, policymakers
Cultural Changes: Integrating Conservation into Forestry	
Build increased public support for private forest conservation and stewardship	Policymakers, public, the media
Reach out strongly to the public to tell the story of stewardship forestry	Public, policymakers, the media
Disseminate forest owner stewardship success stories	Public, policymakers, forest owners and advisors, the media
Expand research and education regarding the ecology and economics of stewardship forestry	Owners, advisors, policymakers, universities, forestry/conservation organizations
Create and support a network of stewardship forestry research and demonstration forests	Public, policymakers, forest owners and advisors

Table 7-1. *(continued)*

Better identify and serve the conservation and stewartship needs of larger forest owners	Larger forest owners
Create new coalitions and partnerships of forest conservation and stewardship allies	Forest owners, influencers, stewardship and conservation organizations, communities, public agencies, and the like
Invest in local and regional conservation and stewardship organizational capacity	Conservation organizations, forest-owner organizations, public agencies, media
Encourage new stewardship service partnerships to assist landowners	Conservation organizations, forest-owner organizations, public agencies
Improve the quality and utility of private forest data and research	Public agencies, conservation organizations, universities, forest owners, foresters
Expand regional and local conservation planning, mapping, and information initiatives	Communities, conservation organizations, forest owners, public, planners

States by making the cost basis more competitive with that of overseas holdings. Similarly, for many larger family forest ownerships the ability to cash in on conservation rather than development can enable the next generation or other successors to be better stewards of forest resources.

The willingness of private capital to coinvest in conserved, managed forests provides partnered public agencies and philanthropies with the opportunity to conserve much more forestland than they could alone. Although private investors can marshal capital for the commercial value of conserved forests, they require public or charitable investment in the "conservation increment" of the forest value. The recent forest conservation transactions in New England described earlier demonstrate how large-scale conservation transactions can combine public, private, and charitable capital to create new protected forests in both public and private ownership.

Not only can grants targeted to particular projects have significant impacts, philanthropic funds can also be pooled regionally or nationally to potentially increase efficiencies of use within an overarching plan. New mechanisms to facilitate the acquisition, conservation, and stewardship of private forests are also needed to organize public, private, and philanthropic capital at appropriate scales. This can occur through the creation of stewardship forestry investment funds managed by forest investment management organizations (FIMOs) or other innovative ownership forms, as described in The Conservation Toolbox under Creating New Steward-ship Ownership Forms.

Philanthropies can use grants and program-related investments (PRIs) in the form of low-cost loans to expand the internal working capital avail-able to nonprofit organizations for forest conservation. As described ear-lier, such revolving conservation funds are instrumental in facilitating con-servation transactions by allowing the conservation organization to acquire or otherwise tie up the property and then organize its long-term conservation and resale to a conservation buyer.

Expand the public market for forest conservation. As the tide of support for antisprawl measures and other open-space programs grows, forest stake-holders and the concerned public need to ensure that funding programs include forest conservation goals. Funding for forest conservation needs to be at least on par with funding for agricultural protection.

There are also great opportunities to build funding and overall public support for forest conservation and stewardship at all levels of govern-ment. Private forest conservationists, philanthropies, and others need to support expansion of federal funding through Forest Legacy, the Land and Water Conservation Fund, and similar mechanisms, as well as through other programs. In addition, greater support is needed to enact state and local taxes or bond financing for open space, targeting private forests man-aged for conservation values. As booming economies and urban popula-tions move increasingly to forest areas, there is a real opportunity to work with interested communities to sponsor more conservation-oriented tax-funding initiatives.

Catalyze new sources of ongoing conservation funding through market devel-opment for forest ecosystem services. When forest conservation becomes more competitive with alternative uses and can provide new revenue streams for stewardship, more landowners will incorporate conservation into their management. Therefore it is important to expand the forest con-servation marketplace beyond traditional funding mechanisms. In partic-ular, the ecosystem services of carbon sequestration and watersheds hold significant potential for economic returns that are compatible with both conservation and stewardship-oriented timber management. In fact rev-enue from ecosystem services can be utilized to acquire CEs on managed forestland to secure the long-term provision of those services, as described under Developing Markets for Ecosystem Services in The Conservation Toolbox. Governmental policy and philanthropies can play important roles in building these new markets, which will then operate primarily on their own.

Forest carbon markets. Although it has only recently emerged, forest carbon marketing has the potential to mobilize substantial conservation funding from new private sources, with estimated future returns in the United States of tens to hundreds of millions of dollars annually. The carbon market today is in its first stage. At least three brokerage firms—Natsource, Cantor Fitzgerald, and Environmental Financial Products—are trading "carbon credits" and options. As of the end of 2000, only one transaction had occurred based on conservation and sustainable management of private forests in the United States, in which PFT sold credits to Green Mountain Energy. Roughly ten transactions of any nature occur annually worldwide. It is still a very small, highly variable, and risk-prone market owing to the evolving policy context, which tends to discourage private investment on any major scale. To diminish this risk internationally, the World Bank has launched its Prototype Carbon Fund with the aim of pioneering standardized transactions and encouraging markets in the developing world.

Catalytic domestic charitable or even public investments could provide the risk capital essential to generating model transactions to help shape the future structure of the domestic market. This capital, in the range of $10 to $50 million, could provide the seed money to establish forest carbon banks dealing in high-quality carbon credits from forestlands secured through CEs, following the model created in PFT's Green Mountain Energy transaction. The easements would ensure permanent stores of more carbon than would otherwise accumulate while preventing its liquidation from unsustainable harvest or conversion. These credits would be resold when the market has grown, recycling the original risk capital into new conservation transactions.

The public and public agencies need to better understand how a well-designed carbon market could benefit forest conservation and stewardship. Their support is crucial to ensuring that the policy framework for such a market is one that genuinely provides incentives for conservation, and not a new subsidy for business as usual.

Forest water market. Water is another major ecosystem service that could provide increased funding for forest conservation. Although water is not yet traded as a commodity per se, a number of conservation transactions detailed in the Toolbox have been effected specifically to protect water supplies. These nascent markets tend to be local or regional in nature, focused on certain municipal watersheds and driven by public agencies. To accelerate conservation of significant private forests in these watersheds, philanthropies and public agencies could support pilot projects in key communities that would enable them to dedicate a portion of their water-use rates to a fund for easement acquisition on private forest ownerships. Development of such a fund would build on existing programs and efforts, such as those described in the Toolbox. One variation would organize the acquisition and management of watershed forestlands by community-based nonprofit organizations, funded through rate-payer mechanisms.

Improve returns from long-term stewardship and conservation of forestlands through changes in key areas of taxation. As outlined in the Public Policy Initiatives portion of the Toolbox, an effective, broad-based coalition comprising the range of forest interests is needed to promote key changes in tax policy as fiscal incentives for forest conservation. Such actions will help fund direct forest conservation, reduce forest fragmentation, and assist landowners in making investments in the restoration of ecosystem wealth. The changes include:

- Providing favorable capital gains tax treatment for conservation transactions and for truly long-term investments (e.g., more than twenty years) in forests
- Enabling donors of conservation easements to more fully realize the value of their charitable gifts either through increased income tax deductions or credits
- Improving the ability of all forest owners to more quickly recover the costs of forest stewardship investments through expensing or accelerated amortization
- Expanding existing estate tax benefits for CEs, and otherwise reducing the impact of estate taxes on private forests as long as they remain as forests

- Improving local property tax systems to promote further forest conservation and create disincentives to development

Increase access to liquidity and traditional sources of capital. As described in chapter 3, working capital for stewardship and conservation is often in critically short supply for small- and mid-sized landowners. This situation needs to be improved to enable the majority of private forest owners to make investments in restoring forest ecosystem wealth, reassembling the landscape, and directly conserving their properties.

Philanthropies and the public sector can encourage these private investments in several important ways. They can make grants and low-cost loans available to forest owners for investments in conservation and stewardship, through existing community economic development organizations, conservation organizations, or other entities earmarked for low-cost lending. They can support state agency efforts to create lending mechanisms. They can also work with nonprofit organizations to develop innovative landowner financing strategies, such as stewardship forest leases, a nonprofit timber futures market, and similar mechanisms.

Increase returns from managing forests with high biodiversity values. Market forces shape forests. Therefore to ensure that more forests are managed for biodiversity, appropriate markets must be available for products harvested from such forests. Landowners who manage for older, higher-value forests in particular need to feel that it is worth the lengthy investment. They also need to feel reasonably assured that there will be a domestic market for large sawtimber in the future. Local and regional markets need to be developed for both small-dimension and large-dimension logs harvested as part of a stewardship forestry plan. Philanthropies and public agencies can help in a number of ways, such as by underwriting research, publications, and conferences on economic returns and markets for value-added, high-quality wood products; and by supporting the organization of landowners and processors who are committed to managing older forests and producing high-quality timber to share in market development. In addition, landowners can be organized to share stewardship costs and the added returns possible through their own collective log merchandising and value-added processing. The general public and forest-product buyers can be encouraged to support well-managed forests through cooperative efforts among forest owners and philanthropies to promote purchases of certified forest products.

Cultural Changes: Integrating Conservation into Forestry and Forestry into Society

Two great cultural barriers must be overcome to achieve the conservation goals laid out here. First, people think that conservation is something that happens separate from forest management. Second, both the forest community and the general public share a feeling that forests are separate and foreign from mainstream urban America. All efforts to expand private forest conservation are hampered by lack of integration of conservation into forestry, and forestry into society.

This section describes culturally directed actions to advance conservation objectives. These actions focus on increased communication and education, and are rooted in expanding essential research. They are generally focused on people rather than on transactions. Their overall objective is to help build a culture that places a higher value on forests. Further, by implementing these activities forest owners will be better engines of conservation who invest in restoring ecosystem wealth, participate in reassembling the forest landscape, and understand the benefits of conservation-based forestry.

Tell the story of good forestry and its financial and ecological returns. Foresters and conservationists need to better define "good," stewardship-oriented forestry and express its private and public benefits. In so doing, they will educate their peers and forest landowners and build support for forestry as an acceptable land use. To accomplish this goal, the conservation and traditional forestry community should cooperate in research and development of case studies that will quantify the costs and benefits of forest conservation and stewardship, with the funding support of philanthropies. The business case for stewardship forestry needs to be made compellingly to compete with the conventional wisdom that conventional forestry is the only economically viable approach. This approach includes, for example, documenting the returns of forestry and wildlife management in the Mississippi Valley; the profit contribution of special forest-products harvesting and processing in Washington State; or the tax benefits to forest-owning families of CE donations.

In addition, more forest landowners need to share their stewardship success stories on a peer basis through publications, field trips to conserved, managed forests, and other cooperative educational efforts among forest landowner associations, conservation and stewardship organizations, foresters, and landowners themselves. Beyond this, a network of demonstration stewardship forests should be created to provide powerful, on-the-ground educational opportunities for landowners, foresters, and the general public to help them better comprehend the practical hows and whys of conservation-based forestry. Drawing from current forests managed by private landowners, conservation organizations, universities, and others, a national network could comprise forests of various types, scales, and stewardship situations.

The public's negative perception of forestry as a land use is itself a barrier to forest conservation and stewardship. As discussed in chapter 3, public distrust of forestry limits public investment in the conservation and stewardship of private, managed forests. Yet many private forestry uses are socially beneficial and can be ecologically beneficial as well. To foster public understanding of stewardship-oriented forestry as a desired land use, organizations should promote examples of excellent forestry at various scales and in various contexts, humanizing the people involved in forestry and graphically illustrating the public benefits of private forests. Forest certification programs can play an important role in building new public respect and appreciation for private forestry. All these efforts will require wider general public outreach through well-designed media campaigns, the use of demonstration forests, and the development of new coalitions.

The environmental community is a key influencer of public opinion regarding forests and forestry. Any effort to address the negative perception of forestry needs to include education, outreach, and cooperation with environmental activists at the local, regional, and national levels so that environmentalists can become comfortable with and supportive of the protection of managed, private forests. One example is the work of the Forest Stewardship Council's forest certification program. Another is the educational work of the Northern Forest Alliance.

Build support among the public and policymakers for increased public investment in private forests. It is essential for forest stakeholders to articulate a compelling case for public investment in the conservation and stewardship of private U.S. forests. This case needs to illustrate current conditions, elucidate the consequences of current trends for biodiversity and other critical forest values, and demonstrate the need for concerted effort by government at all levels to increase coinvestment with forest owners to benefit public trust values in private forests. Regional, state, and national versions of this case would be useful. Public outreach to communicate it, using various media and venues, is essential. Without better public understanding and support, many strategically important actions will be hampered, such as new market development, taxation, and financing mechanisms.

Improve identification and understanding of key forestland owners to provide them with more effective conservation and stewardship services. The large number of forestland owners itself makes outreach to them both overwhelming and traditionally ineffective. Information on who they are, why they own forestland, and what their stewardship challenges are is very limited, if it exists at all. Public agencies, foresters, conservationists, and other stewardship service providers need better information about forest owners to provide more extensive, more relevant services to them. Efforts should focus on understanding the perceptions and better serving the needs of larger landowners who control 45% of forestlands in ownerships of 500 acres or more. Identifying those top forest owners provides the opportunity to effectively concentrate energies on fewer entities, but with greater extent of forest. Building a better database entails contacting, polling, and holding focus groups for these forest landowners to determine how to form new coalitions, create more effective programs using existing tools, and, as needed, design new tools that these owners will use to increase conservation results. Special attention needs to be paid to identifying and addressing landowners' perceptions of barriers to conservation and stewardship at regional and local levels. We also need to gain a better understanding of the forest succession issues family ownerships face so as to better craft assistance in maintaining them through time.

Convene and build new coalitions and partnerships among natural forest conservation allies. There is growing recognition that conservation and forestry interests share a profound concern for the protection and stewardship of private forests that pulls them together despite other issues that may divide them. Past efforts to create new partnerships, however, have suffered from being both too narrow—just within an interest sector—or too broad and insufficiently focused. Sometimes these efforts have been too limited in time to accomplish much. We now recognize that the broad range of forest stakeholders—including conservationists, forest owners and managers, "influencers," urban and rural communities, public-sector agencies, and others—needs to be brought together to build new coalitions for public support of private forest conservation at local, state, regional, and national levels. This step is fundamental to ensuing leadership not only to expand public funding mechanisms but to promote wider public engagement with forests. Such new alliances need a long-term horizon to build lasting partnerships appropriate to the special challenges of forest conservation. Key to this effort is the philanthropic support of local, regional, and national organizations, and of individual "civic entrepreneurs" capable of acting as catalysts and conveners for the pursuit of the private-forest conservation agenda.

In a related strategic approach, greater linkages need to be forged among various stewardship-oriented practitioners and organizations. Community forestry enterprises, forester networks, landowner associations, certification groups, land trusts, environmental justice activists, transportation activists, and others can fruitfully interact and cross-fertilize their initiatives, strengthening the bonds among them.

Enhance regional stewardship capacity to support landowners in conservation. Whether directed at larger or smaller forest properties, many of these increased conservation activities are going to be complex and of a significant scale within distinct regions. They will require a sophisticated level of capacity to ensure that conservation transactions and stewardship programs are properly implemented and maintained over time. Within the forest conservation community, this expertise and capacity needs further support and development. Collaborations among service providers—such as land trusts, forest landowner associations, university extension, and Resource Conservation Districts—should be encouraged to expand the resources and reach of programs. Such stewardship service partnerships would deliver more and better assistance to forest owners through enhanced convenience and one-stop shopping for the range of services provided by nonprofits and government agencies. In regions such as the South, creation of new forest conservation organizations may be needed to provide expertise and leadership in forest conservation, stewardship, education, and outreach.

Although there is depth at the national level and within certain regions, individuals, and organizations, forest conservation organizations in general must improve their financial and technical capacities to meet the growing demand for their services. They face high hurdles in negotiating large, complex transactions with multiple partners, and in meeting the long-term stewardship obligations of conserved managed forests. In addition to philanthropic investment in conservation capacity building, national conservation organizations with greater financial means and valuable relationships should be encouraged to support and form joint ventures with local or regional organizations. The latter can provide knowledge of a particular region and its resources and contribute other skills.

Improve access to scientific information to advance forest conservation and stewardship. As many will testify, data on private-forest resources and conditions available from federal and state agencies are spotty, often inconsistent or incompatible in form or definition, poorly integrated, and frequently out of date. Lack of good current information—especially that which illustrates conditions and trends in nontimber resources—is itself a threat to private-forest conservation and stewardship. Improving these data, as well as useful analyses, is a key objective that must be accomplished if we are to meet our strategic goals.

It is outside the scope of this book to propose specific changes or enhancements to the data collection and research efforts of government agencies; however, the National Research Council and others have done so (1998). Those concerned with conservation of private forests can help improve the availability, accuracy, and utility of publicly funded research on private forests in a number of ways. They should make the need for improvement very clear to policymakers. They also should support efforts to assess research requirements and data for forest conservation. In particular, a national research coordinating council that includes conservation interests should be organized. Local, regional, and national analyses of forest conditions, threats, and conservation opportunities should be organized and funded. In addition, these analyses should be packaged and disseminated so that the numbers come alive and mobilize increased efforts. Greater participation by forest conservationists, foresters, and landowners in the results of the Montreal Process, including the work of the U.S. Roundtable on Sustainable Forests, is also needed. This will ensure that the effort to create criteria and indicators of sustainability is thoroughly reviewed by those who will be using them. Further, it is good to remember that research follows the money, like everything else. Philanthropies and government agencies should expand university and other funding for research related to the conservation of biodiversity and the restoration of robust ecological functioning on managed private forests.

Forest communities and regions are often generally aware that some of their resources may be at risk, but lacking a clear picture of which resources and the degrees and consequences of those risks, they do not act. At the same time, a rapidly increasing amount of data is available, often in digital form. Unfortunately, communities do not have access or equipment to use such data. Nonprofit organizations and government agencies should prepare and provide compatible, accessible data banks and mapping services for local communities to help them identify conservation opportunities and needs for their regions.

Demonstration forests are key tools in forest conservation. They provide both on-the-ground laboratories for developing applied stewardship forestry as well as classrooms that enable landowners, foresters, communities, policymakers, and the general public to comprehend ecologically beneficial forestry. As mentioned earlier, existing research and demonstration forests could be organized into networks similar to the National Science Foundation's network of Long-Term Ecological Research sites. Demonstration forests can be managed to provide information about stewardship management, including silvicultural approaches and practical ecological monitoring systems. Revenue generated from these forests would support their operating costs, while philanthropies could provide grants to fund specific research projects.

Conclusion

This conservation strategy focuses on accomplishing key goals through actions that generate both immediate, direct gains and long-term reduction of the threats to private forests. It lays out a broad scheme of action. Various organizations, agencies, and individuals will take up different pieces of the strategy. Some may seek to implement it holistically but will do so at a local or regional scale. Although engaging the power of forest owners to advance conservation and stewardship is a key element for success, this are not the only group crucial to the conservation of America's threatened private forests. All circles of forest stakeholders and society at large must become more engaged. As they do, and as this strategy is implemented by those concerned for the future of private forests, everyone will learn a great deal about how better to accomplish these vital conservation goals. It is the aspiration of the authors that this strategy be the starting point for a burgeoning forest conservation movement that continues to evolve. Then perhaps at the end of the new century, people will be able to enjoy privately owned forests of the same extent—and of even greater quality—as those they enjoy today.

Appendix A.
List of Acronyms

BIA: Bureau of Indian Affairs

BMP: Best Management Practices program, a state program (either voluntary or regulatory) encouraged by the EPA since the Clean Water Act to control negative impacts to water quality

CCC: Civilian Conservation Corps, established by President Roosevelt in the 1930s to put people to work planting trees, stopping gullies, fighting forest fires, and building dams

CE: conservation easement

CSREES: United States Department of Agriculture's Cooperative State Research, Education and Extension Service

ESA: Endangered Species Act

EQIP: Environmental Quality Incentives Program, created in the 1996 Farm Bill to provide farmers and ranchers with educational, financial, and technical assistance to address threats to soil, water, and other natural resources

FIA: Forest Inventory and Analysis, a program of the USDA Forest Service

FIMO: Forest Investment Management Organization, proposed by the Pacific Forest Trust to be modeled after TIMOs. FIMOs can organize funds of investors, or manage dedicated accounts for large institutions, to provide competitive risk-adjusted returns through the acquisition, conservation, and management of forestland on the stewardship forestry business model.

FSA: Farm Services Agency

FSC: Forest Stewardship Council, an international body that oversees forest certification

FSP: Forest Stewardship Program, created in the 1990 Farm Bill

HCPs: Habitat Conservation Plans, established under Section 10 of the

ESA. These plans, submitted by nonfederal landowners, are intended to minimize and mitigate the impacts of the approved management activities, allowing for a limited or incidental degree of harm to species in return for a commitment on the part of the landowner to maintain or enhance habitat overall, in accordance with the plan.

IPFs: Industrial forest landowners are those who also own and operate wood-processing facilities.

LAPs: Landowner Assistance Programs, established by forest product companies to assist NIPFs in forest management planning. These programs effectively provide private-sector forestry assistance to supplement extension foresters.

MLPs: master limited partnerships

NEPA: National Environmental Policy Act, passed in 1969 and focused on the need to consider the broad environmental impact of major public actions

NIPFs: Nonindustrial forest landowners include all other landowners, large and small, in all forms of ownership.

NMFS: National Marine Fisheries Service

NRC: National Research Council

NRCS: Natural Resources Conservation Service

NRI: National Resources Inventory, a report produced every five years by the NRCS which tracks a number of factors related to land use

RCDs: Resource Conservation Districts

REIT: real estate investment trusts

RREA: Renewable Resources Extension Act

SFAs: state forestry agencies

SFI: Sustainable Forestry Initiative, operated by the American Forest & Paper Association to certify the sustainability practices of its members

SIP: Stewardship Incentives Program

Appendix B.
Tables

Table B-1.
Forest and timberland ownership in the United States by state and region, 1992

State/Region	Forestland				Timberland				
	All Owners	Total Private	Forest Industry	NIPF	All Owners	Total % Private	Forest-land	Forest Industry	NIPF
				(thousand acres)					
Arizona	19,926	8,214	13	8,201	4,073	1,297	15.7%	0	1,297
Colorado	21,270	5,939	0	5,939	11,555	3,224	54.3%	0	3,224
Idaho	21,937	3,389	1,284	2,106	17,123	3,222	95.0%	1,284	1,938
Montana	23,232	6,267	1,618	4,649	19,164	5,957	95.1%	1,618	4,340
Nevada	9,928	580	25	555	169	82	14.1%	25	57
New Mexico	15,505	6,042	0	6,042	4,833	1,958	32.4%	0	1,958
Utah	15,705	2,744	0	2,744	4,700	878	32.0%	0	878
Wyoming	10,944	1,977	0	1,977	5,085	1,444	73.0%	0	1,444
Inter-mountain	**138,447**	**35,152**	**2,939**	**32,213**	**66,701**	**18,063**	**51.4%**	**2,926**	**15,137**
Kansas	1,545	1,436	0	1,436	1,491	1,399	97.4%	0	1,399
Nebraska	947	814	0	814	898	790	97.1%	0	790
North Dakota	674	442	0	442	442	387	87.6%	0	387
South Dakota	1,632	557	0	557	1,487	485	87.1%	0	485
Great Plains	**4,798**	**3,248**	**0**	**3,248**	**4,317**	**3,062**	**94.3%**	**0**	**3,062**
Illinois	4,294	3,648	13	3,635	4,058	3,641	99.8%	13	3,628
Indiana	4,501	3,731	17	3,713	4,342	3,719	99.7%	17	3,701
Iowa	2,050	1,807	0	1,807	1,944	1,788	98.9%	0	1,788
Michigan	19,335	12,138	1520	10,618	18,667	12,039	99.2%	1,514	10,525
Minnesota	16,796	7290	761	6,529	14,819	7,139	97.9%	751	6,388
Missouri	14,047	11,626	222	11,403	13,411	11,359	97.7%	222	11,137
Ohio	7,855	7,165	174	6,990	7,568	7,036	98.2%	174	6,862
Wisconsin	15,963	11,196	1,105	10,091	15,701	11,155	99.6%	1,102	10,053
North Central	**84,842**	**58,599**	**3,814**	**54,785**	**80,510**	**57,877**	**98.8%**	**3,795**	**54,082**

Table B-1. (*continued*)

	Forestland				Timberland				
	All Owners	Total Private	Forest Industry	NIPF	All Owners	Total % Private	Forest-land	Forest Industry	NIPF
State/Region					(thousand acres)				
Connecticut	1,863	1,591	0	1,591	1,815	1,565	98.4%	0	1,565
Delaware	389	373	31	342	376	363	97.3%	31	332
Maine	17,711	16,732	7,449	9,283	16,952	16,323	97.6%	7,298	9,024
Maryland	2,701	2,267	137	2,130	2,423	2,143	94.5%	137	2,006
Massachusetts	3,264	2,622	71	2,552	2,965	2,486	94.8%	71	2,415
New Hampshire	4,955	3,862	519	3,343	4,551	3,758	97.3%	513	3,246
New Jersey	1,991	1,386	0	1,386	1,864	1,364	98.4%	0	1,364
New York	18,581	14,454	1,225	13,229	15,406	14,252	98.6%	1,220	13,032
Pennsylvania	16,905	12,502	613	11,889	15,853	12,334	98.7%	613	11,721
Rhode Island	409	314	0	314	356	287	91.4%	0	287
Vermont	4,607	3,880	227	3,653	4,461	3,868	99.7%	227	3,642
West Virginia	12,108	10,588	887	9,701	11,900	10,576	99.9%	887	9,689
Northeast	**85,484**	**70,570**	**11,158**	**59,412**	**78,923**	**69,320**	**98.2%**	**10,996**	**58,324**
Alaska	**127,380**	**35,875**	**0**	**35,875**	**12,395**	**3,790**	**10.6%**	**0**	**3.790**
Oregon	29,720	10,775	5,290	5,485	23,749	8,626	80.1%	5,012	3,613
Washington	21,892	9,811	4,305	5,506	17,418	8,954	91.3%	4,109	4,845
Pacific NW	**51,612**	**20,586**	**9,595**	**10,991**	**41,167**	**17,580**	**85.4%**	**9,121**	**8,458**
California	38,547	16,754	3,140	13,613	17,952	7,437	44.4%	2,982	4,455
Hawaii	1,748	1,155	0	1,155	700	362	31.3%	0	362
Pacific SW	**40,296**	**17,909**	**3,140**	**14,768**	**18,652**	**7,798**	**43.5%**	**2,982**	**4,816**
Alabama	21,964	20,781	4,796	15,985	21,911	20,781	100.0%	4,796	15,985
Arkansas	18,790	15,260	4,498	10,762	18,392	15,118	99.1%	4,498	10,620
Kentucky	12,684	11,368	205	11,164	12,347	11,344	99.8%	205	11,139
Louisiana	13,783	12,479	3,899	8,579	13,693	12,479	100.0%	3,899	8,579
Mississippi	18,595	16,651	3,241	13,411	18,587	16,651	100.0%	3,241	13,411
Oklahoma	7,665	7,008	1,049	5,959	6,234	5,659	80.8%	1,049	4,610
Tennessee	13,603	11,762	1,122	10,641	13,265	11,757	99.9%	1,122	10,635
Texas	18,354	17,446	3,720	13,726	11,766	10,990	63.0%	3,720	7,271
South Central	**125,438**	**112,755**	**22,529**	**90,226**	**116,196**	**104,778**	**93.0%**	**22,529**	**82,249**
Florida	16,254	12,158	4,018	8,140	14,605	11,819	97.0%	4,016	7,803
Georgia	24,413	22,048	4,381	17,667	23,796	22,045	99.9%	4,381	17,664
North Carolina	19,298	16,774	2,252	14,522	18,639	16,760	99.9%	2,252	14,508
South Carolina	12,651	11,341	2,322	9,019	12,419	11,341	100.0%	2,322	9,019
Virginia	25,430	13,466	1,537	11,929	15,345	13,465	99.9%	1,537	11,927
Southeast	**88,662**	**75,788**	**14,511**	**61,277**	**84,803**	**75,430**	**99.5%**	**14,508**	**60,922**
Total U.S.	**746,958**	**430,483**	**67,687**	**362,796**	**503,664**	**357,698**	**67.4%**	**66,858**	**290,840**
% of Total	**100%**	**58%**	**9%**	**49%**	**100%**	**71%**		**13%**	**58%**

Source: Compiled from U.S. Forest Service data by the Sampson Group.

Table B-2.

Individual private owners aged 65 years or More by U.S. Region

Region	Individuals 65 yrs+	% Individuals	Acreage	% Acreage
Pacific Coast	177,500.00	2	6,840.00	2
Intermountain	52,400.00	1	3,149.00	1
Great Plains	63,800.00	1	1,297.00	0
North Central	545,800.00	6	18,235.00	5
Northeast	549,000.00	6	15,869.00	4
Southeast	464,600.00	5	18,758.00	5
South Central	691,000.00	7	28,511.00	7
65 yrs+ total	2,544,100.00	26	92,659.00	24

Source: Birch 1996

Tables B-3.

Analysis of Birch 1996 ownership data by region: Pacific Coast

Profile of Pacific Coast private forestland ownership by owners and acres owned

	1–9 acres	10–99 acres	100–499 acres	500–999 acres	1,000+ acres	Total
Landowners	395,300	211,300	31,100	3,500	3,300	644,500
% total owners	61%	33%	5%	1%	1%	100%
Forest acres	1,123,000	6,525,000	5,371,000	2,193,000	30,614,000	45,826,000
% total acres	2%	14%	12%	5%	67%	100%

Pacific Coast forest owners by type of entity

Ownership Type	Ownerships	% Ownership	Acreage	% Acreage	Avg. Acreage
Forest industry	1,000	0%	14,529,000	39%	14,529
Farm	98,900	15%	8,568,000	23%	87
Industrial business	3,900	1%	972,000	3%	249
Real estate	9,400	1%	1,246,000	3%	133
Nonindustrial business *	8,325	1%	193,000	1%	23
Recreation club	17,500	3%	491,000	1%	28
Public utility *	25	0%	113,000	0%	4,520
Individuals	482,000	75%	462,000	1%	1
Other	23,500	4%	10,887,000	29%	463
Total	**644,550**	**100%**	**37,461,000**	**100%**	**58**

* Estimated ownerships with an error of plus or minus 25

Tables B-3. (*continued*)

Primary Reason For Owning Forestland for the Pacific Coast

Reason	Ownerships	% Owners	Acres	% Acres
Timber	13,600	2%	17,967,000	41%
Land investment	77,900	14%	2,804,000	6%
Farm and domestic uses	21,100	4%	2,871,000	7%
Part of residence	136,100	24%	2,470,000	6%
Part of farm	56,700	10%	2,548,000	6%
Recreation	52,400	9%	2,705,000	6%
Enjoyment	129,100	23%	2,047,000	5%
Other	80,500	14%	9,268,000	21%
No answer	3,800	1%	1,223,000	3%
Total	**571,200**	**100%**	**43,903,000**	**100%**

Figures do not include estates

Pacific Forestland Ownership by Date of Acquisition

Years	Total Owners	% Owners	Acres	% Acres
1990–1994	122,200	19%	1,668,000	4%
1980–1989	206,500	32%	6,290,000	14%
1970–1979	127,200	20%	10,791,000	24%
1960–1969	69,100	11%	3,201,000	7%
1950–1959	67,400	10%	5,099,000	11%
1940–1949	18,900	3%	3,119,000	7%
1901–1939	7,000	1%	4,021,000	9%
Before 1900	1,600	0%	10,410,000	23%
No answer	24,500	4%	1,231,000	3%
Total	**644,400**	**100%**	**45,830,000**	**100%**

Pacific Forestland Ownership by Occupation

Occupation	Total Owners	% Owners	Acres	% Acres
White collar	231,400.00	37%	4,290,000	10%
Blue collar	94,200.00	15%	1,173,000	3%
Farmer	26,000.00	4%	2,791,000	6%
Service worker	16,100.00	3%	121,000	0%
Homemaker	2,900.00	0%	108,000	0%
Retired	192,900.00	31%	5,290,000	12%
Other	63,400.00	10%	31,000,000	69%
Total	**626,900.00**	**100%**	**44,773,000**	**100%**

Figures do not include landowners who did not respond

Ownership Expectations of Future Timber Harvest for the Pacific

	Expected Future Harvest	% Owners % Acres
1–10 years	37%	78%
Indefinite	22%	12%
Never	36%	9%
No answer	6%	1%

Tables B-4.

Analysis of Birch 1996 ownership data by region: Intermountain

Profile of Intermountain Private Forestland Ownership by Owners and Acres Owned

	1–9 acres	10–99 acres	100–499 acres	500–999 acres	1,000+ acres	Total
Landowners	138,880	109,000	19,200	2,300	2,800	272,180
% total owners	51%	40%	7%	1%	1%	100%
Forest acres	642,000	2,670,000	3,211,000	1,554,000	19,339,000	27,416,000
% total acres	2%	10%	12%	6%	71%	100%

Intermountain Forest Owners by Type of Entity

Ownership Type	Ownerships	% Ownership	Acreage	% Acreage	Avg. Acreage
Forest industry *	125	0%	9,424,000	34%	75,392
Farm	68,000	25%	9,987,000	36%	147
Industrial business	1,100	0%	317,000	1%	288
Real estate	3,600	1%	517,000	2%	144
Nonindustrial business	100	0%	51,000	0%	510
Recreation club	3,700	1%	541,000	2%	146
Public utility	—	0%	—	0%	—
Individuals	190,600	70%	2,700,000	10%	14
Other	5,100	2%	3,881,000	14%	761
Total	**272,325**	**100%**	**27,418,000**	**100%**	**101**

* Estimated ownerships with an error of plus or minus 25

Tables B-4. (*continued*)

Primary Reason for Owning Forestland for the Intermountain

Reason	Ownerships	% Owners	Acres	% Acres
Timber	400	0%	4,986,000	19%
Land investment	12,800	5%	1,157,000	4%
Farm and domestic uses	18,000	7%	6,436,000	24%
Part of residence	43,300	17%	1,249,000	5%
Part of farm	16,200	6%	2,785,000	10%
Recreation	44,500	18%	2,486,000	9%
Enjoyment	67,300	27%	1,426,000	5%
Other	49,000	19%	5,681,000	21%
No answer	600	0%	682,000	3%
Total	**252,100**	**100%**	**26,888,000**	**100%**

Figures do not include estates

Intermountain Forestland Ownership by Date of Acquisition

Years	Total Owners	% Owners	Acres %	Acres
1990–1994	51,700	19%	1,543,000	6%
1980–1989	43,700	16%	2,855,000	10%
1970–1979	109,900	40%	2,657,000	10%
1960–1969	22,500	8%	2,493,000	9%
1950–1959	12,600	5%	1,536,000	6%
1940–1949	11,300	4%	1,440,000	5%
1901–1939	13,100	5%	2,561,000	9%
Before 1900	500	0%	11,799,000	43%
No answer	6,900	3%	533,000	2%
Total	**272,200**	**100%**	**27,417,000**	**100%**

Intermountain Forestland Ownership by Occupation

Occupation	Total Owners	% Owners	Acres	% Acres
White collar	120,500	44%	2,428,000	9%
Blue collar	26,900	10%	307,000	1%
Farmer	38,900	14%	3,461,000	13%
Service worker	14,600	5%	136,000	1%
Homemaker	3,400	1%	112,000	0%
Retired	45,900	17%	1,815,000	7%
Other	21,200	8%	18,926,000	70%
Total	**271,400**	**100%**	**27,185,000**	**100%**

Figures do not include landowners who did not respond

Ownership Expectations of Future Timber Harvest for the Intermountain

Expected Future Harvest	% Owners	% Acres
1–10 years	26%	75%
Indefinite	20%	11%
Never	53%	13%
No answer	0%	1%

Tables B-5.

Analysis of Birch 1996 ownership data by region: Plains

Profile of The Plains: Private Forestland Ownership by Owners and Acres Owned

	1–9 acres	10–99 acres	100–499 acres	500–999 acres	1,000+ acres	Total
Landowners	66,500.00	40,500.00	5,800.00	600.00	100.00	113,500.00
% total owners	59%	36%	5%	1%	0%	100%
Forest acres	182,000	1,336,000	890,000	374,000	155,000	2,937,000
% total acres	6%	45%	30%	13%	5%	100%

Plains Forest Owners by Type of Entity

Ownership Type	Ownerships	% Ownership	Acreage	% Acreage	Avg. Acreage
Forest industry *	25	0%	36,000	1%	1,440
Farm	92,200	81%	2,301,000	78%	25
Industrial business	500	0%	36,000	1%	72
Real estate	—	0%	—	0%	—
Nonindustrial business	—	0%	—	0%	—
Recreation club	1,000	1%	97,000	3%	97
Public utility	—	0%	—	0%	—
Individuals	19,200	17%	407,000	14%	21
Other	700	1%	60,000	2%	86
Total	**113,625**	**100%**	**2,937,000**	**100%**	**26**

* Estimated ownerships with an error of plus or minus 25

Tables B-5. (*continued*)

Primary Reason for Owning Forestland for the Plains

Reason	Ownerships	% Owners	Acres	% Acres
Timber	800	1%	96,000	3%
Land investment	100	0%	36,000	1%
Farm and domestic uses	18,400	16%	494,000	17%
Part of residence	18,000	16%	312,000	11%
Part of farm	61,300	54%	1,242,000	43%
Recreation	7,400	7%	340,000	12%
Enjoyment	2,700	2%	171,000	6%
Other	3,500	3%	144,000	5%
No answer	1,300	1%	74,000	3%
Total	113,500	100%	2,909,000	100%

Figures do not include estates

Plains Forestland Ownership by Date of Acquisition

Years	Total Owners	% Owners	Acres	% Acres
1990–1994	300	0%	23,000	1%
1980–1989	10,900	10%	396,000	13%
1970–1979	35,400	31%	583,000	20%
1960–1969	9,600	8%	594,000	20%
1950–1959	6,800	6%	415,000	14%
1940–1949	39,200	34%	549,000	19%
1901–1939	9,900	9%	212,000	7%
Before 1900	600	1%	86,000	3%
No answer	1,800	2%	78,000	3%
Total	**114,500**	**100%**	**2,936,000**	**100%**

Plains Forestland Ownership by Occupation

Occupation	Total Owners	% Owners	Acres	% Acres
White collar	14,900	15%	294,000	11%
Blue collar	2,800	3%	163,000	6%
Farmer	21,800	22%	729,000	26%
Service worker	—	0%	—	0%
Homemaker	12,300	12%	37,000	1%
Retired	37,400	38%	850,000	31%
Other	9,500	10%	688,000	25%
Total	**98,700**	**100%**	**2,761,000**	**100%**

Figures do not include landowners who did not respond

Ownership Expectations of Future Timber Harvest for the Plains

Expected Future Harvest	% Owners	% Acres
1–10 years	14%	49%
Indefinite	34%	28%
Never	29%	20%
No answer	22%	3%

Tables B-6.

Analysis of Birch 1996 ownership data by region: North Central

Profile of Intermountain Private Forestland Ownership by Owners and Acres Owned.

	1–9 acres	10–99 acres	100–499 acres	500–999 acres	1,000+ acres	Total
Landowners	138,880	109,000	19,200	2,300	2,800	272,180
% total owners	51%	40%	7%	1%	1%	100%
Forest acres	642,000	2,670,000	3,211,000	1,554,000	19,339,000	27,416,000
% total acres	2%	10%	12%	6%	71%	100%

North Central Forest Owners by Type of Entity

Ownership Type	Ownerships	% Ownership	Acreage	% Acreage	Avg. Acreage
Forest industry	1,000	0%	4,791,000	8%	4,791
Farm	709,300	42%	25,015,000	43%	35
Industrial business	10,200	1%	1,075,000	2%	105
Real Estate	35,500	2%	1,463,000	3%	41
Nonindustrial business *	0,825	1%	385,000	1%	36
Recreation club **	36,500	2%	1,663,000	3%	46
Public utility **	125	0%	586,000	1%	4,688
Individuals	848,900	50%	21,130,000	36%	25
Other	31,200	2%	2,139,000	4%	69
Total	1,683,550	100%	58,247,000	100%	35

* Estimated ownerships with an error of plus or minus 25
** Estimated acreage with an error of plus or minus 25 (thousands)

Tables B-6. (*continued*)

Primary Reason for Owning Forestland for the North Central

Reason	Ownerships	% Owners	Acres	% Acres
Timber	17,900	1%	7,084,000	13%
Land investment	76,800	5%	3,189,000	6%
Farm and domestic uses	188,500	12%	7,666,000	14%
Part of residence	357,000	22%	6,024,000	11%
Part of farm	394,900	24%	10,373,000	19%
Recreation	267,400	17%	11,462,000	21%
Enjoyment	229,700	14%	5,604,000	10%
Other	49,100	3%	3,405,000	6%
No answer	31,900	2%	757,000	1%
Total	**1,613,200**	**100%**	**55,564,000**	**100%**

North Central Forestland Ownership by Date of Acquisition

Years	Total Owners	% Owners	Acres	% Acres
1990–1994	79,400	5%	2,169,000	4%
1980–1989	314,500	19%	8,122,000	14%
1970–1979	458,600	27%	14,277,000	24%
1960–1969	442,900	26%	12,018,000	21%
1950–1959	156,100	9%	7,999,000	14%
1940–1949	97,800	6%	4,129,000	7%
1901–1939	46,700	3%	4,309,000	7%
Before 1900	9,900	1%	1,903,000	3%
No answer	78,000	5%	3,368,000	6%
Total	**1,683,900**	**100%**	**58,294,000**	**100%**

North Central Forestland Ownership by Occupation

Occupation	Total Owners	% Owners	Acres	% Acres
White collar	401,900	25%	11,099,000	20%
Blue collar	265,200	17%	5,584,000	10%
Farmer	217,900	14%	9,488,000	17%
Service worker	42,000	3%	1,228,000	2%
Homemaker	30,600	2%	864,000	2%
Retired	521,500	33%	14,567,000	26%
Other	106,000	7%	12,446,000	23%
Total	**1,585,100**	**100%**	**55,276,000**	**100%**

Figures do not include landowners who did not respond

Ownership Expectations of Future Timber Harvest for the North Central

Expected Future Harvest	% Owners	% Acres
1–10 years	39%	61%
Indefinite	29%	24%
Never	26%	13%
No answer	7%	2%

Tables B-7.

Analysis of Birch 1996 ownership data by region: Northeast

Profile of Northeast Private Forestland Ownership by Owners and Acres Owned

	1–9 acres	10–99 acres	100–499 acres	500–999 acres	1,000+ acres	Total
Landowners	1,342,000	768,900	127,100	6,300	3,300	2,247,600
% total owners	60%	34%	6%	0%	0%	100%
Forest acres	3,680,000	23,766,000	19,257,000	3,702,000	20,852,000	71,257,000
% Total Acres	5%	33%	27%	5%	29%	100%

Primary Reason for Owning Forestland for the Northeast

Reason	Ownerships	% Owners	Acres	% Acres
Timber	20,200	1%	17,061,000	27%
Land investment	138,600	8%	2,197,000	3%
Farm and domestic uses	195,400	11%	6,542,000	10%
Part of residence	357,000	20%	7,696,000	12%
Part of farm	204,600	11%	5,801,000	9%
Recreation	248,400	14%	8,468,000	13%
Enjoyment	405,100	22%	7,846,000	12%
Other	119,800	7%	6,465,000	10%
No answer	123,600	7%	1,051,000	2%
Total	1,812,700	100%	63,127,000	100%

Figures do not include estates

Tables B-7. (*continued*)

Northeast Forest Owners by Type of Entity

Ownership Type	Ownerships	% Ownership	Acreage	% Acreage Acreage	Avg.
Forest industry	7,700	0%	11,670,000	16%	1,516
Farm	383,800	17%	14,303,000	20%	37
Industrial business	2,900	0%	1,932,000	3%	666
Real estate	33,400	1%	3,755,000	5%	112
Nonindustrial business*	20,625	1%	1,418,000	2%	69
Recreation club	24,400	1%	2,100,000	3%	86
Public utility*	525	0%	688,000	1%	1,310
Individuals	1,708,300	76%	30,462,000	43%	18
Other	63,200	3%	4,928,000	7%	78
Total	**2,244,850**	**100%**	**71,256,000**	**100%**	**32**

* Estimated ownerships with an error of plus or minus 25

Northeast Forestland Ownership by Date of Acquisition

Years	Total Owners	% Owners	Acres	% Acres
1990–1994	211,900	9%	1,856,000	3%
1980–1989	541,400	24%	12,556,000	18%
1970–1979	477,300	21%	12,663,000	18%
1960–1969	391,800	17%	11,453,000	16%
1950–1959	166,100	7%	7,400,000	10%
1940–1949	137,800	6%	6,553,000	9%
1901–1939	90,000	4%	6,704,000	9%
Before 1900	14,300	1%	8,789,000	12%
No answer	217,100	10%	3,282,000	5%
Total	**2,247,700**	**100%**	**71,256,000**	**100%**

Northeast Forestland Ownership by Occupation

Occupation	Total Owners	% Owners	Acres	% Acres
White collar	761,400	38%	14,073,000	21%
Blue collar	334,900	17%	5,422,000	8%
Farmer	112,500	6%	5,069,000	8%
Service worker	34,100	2%	853,000	1%
Homemaker	33,100	2%	1,263,000	2%
Retired	618,200	31%	14,975,000	22%
Other	122,000	6%	25,277,000	38%
Total	**2,016,200**	**100%**	**66,932,000**	**100%**

Figures do not include landowners who did not respond

Ownership Expectations of Future Timber Harvest for the Northeast

Expected Future Harvest	% Owners	% Acres
1–10 years	31%	62%
Indefinite	29%	24%
Never	32%	10%
No answer	8%	4%

Tables B-8.
Analysis of Birch 1996 ownership data by region: Southeast

Profile of Southeast Private Forestland Ownership by Owners and Acres Owned

	1–9 acres	10–99 acres	100–499 acres	500–999 acres	1,000+ acres	Total
Landowners	1,726,600	590,600	105,300	10,900	7,700	2,441,100
% total owners	71%	24%	4%	0%	0%	100%
Forest acres	4,130,000	18,347,000	17,300,000	6,410,000	30,132,000	76,319,000
% total acres	5%	24%	23%	8%	39%	100%

Southeast Forest Owners by Type of Entity

Ownership Type	Ownerships	% Ownership	Acreage	% Acreage	Avg. Acreage
Forest industry	1,300	0%	15,821,000	21%	12,170
Farm	351,000	14%	18,951,000	25%	54
Industrial business	6,500	0%	1,604,000	2%	247
Real estate	123,900	5%	4,597,000	6%	37
Nonindustrial business	10,825	0%	385,000	1%	36
Recreation club	6,900	0%	1,134,000	2%	164
Public utility*	25	0%	548,000	1%	21,920
Individuals	1,875,300	77%	28,696,000	38%	15
Other	48,400	2%	3,798,000	5%	78
Total	2,424,150	100%	75,534,000	100%	31

* Estimated ownerships with an error of plus or minus 25

Tables B-8. (*continued*)

Primary Reason for Owning Forestland for the Southeast

Reason	Ownerships	% Owners	Acres	% Acres
Timber	125,900	6%	27,945,000	41%
Land investment	384,600	19%	12,107,000	18%
Farm and domestic uses	110,700	5%	3,401,000	5%
Part of residence	744,300	36%	5,562,000	8%
Part of farm	150,300	7%	5,987,000	9%
Recreation	88,700	4%	4,035,000	6%
Enjoyment	356,500	17%	5,321,000	8%
Other	73,500	4%	3,335,000	5%
No answer	40,900	2%	686,000	1%
Total	**2,075,400**	**100%**	**68,379,000**	**100%**

Figures do not include estates

Southeast Forestland Ownership by Date of Acquisition

Years	Total Owners	% Owners	Acres	% Acres
1990–1994	226,700	9%	3,830,000	5%
1980–1989	785,800	32%	13,913,000	18%
1970–1979	482,500	20%	12,342,000	16%
1960–1969	317,400	13%	11,526,000	15%
1950–1959	168,300	7%	9,054,000	12%
1940–1949	131,700	5%	6,112,000	8%
1901–1939	37,700	2%	11,151,000	15%
Before 1900	35,100	1%	3,316,000	4%
No answer	255,800	10%	5,076,000	7%
Total	**2,441,000**	**100%**	**76,320,000**	**100%**

Southeast Forestland Ownership by Occupation

Occupation	Total Owners	% Owners	Acres	% Acres
White collar	759,600.00	34%	14,527,000	7%
Blue collar	402,300.00	18%	3,956,000	2%
Farmer	113,000.00	5%	5,923,000	3%
Service worker	179,700.00	8%	1,301,000	1%
Homemaker	38,100.00	2%	1,528,000	1%
Retired	573,900.00	26%	15,481,000	8%
Other	142,900.00	6%	161,040,000	79%
Total	**2,209,500.00**	**100%**	**203,756,000**	**100%**

Figures do not include landowners who did not respond

Ownership Expectations of Future Timber Harvest for the Southeast

Expected Future Harvest	% Owners	% Acres
1–10 years	28%	59%
Indefinite	27%	28%
Never	41%	11%
No answer	3%	1%

Tables B-9.

Analysis of Birch 1996 Ownership Data by Region: South Central

Profile of South Central Private Forestland Ownership by Owners and Acres Owned

	1–9 acres	10–99 acres	100–499 acres	500–999 acres	1,000+ acres	Total
Landowners	1,431,500	890,200	155,800	13,400	8,100	2,499,000
% total owners	57%	36%	6%	1%	0%	100%
Forest acres	4,223,000	28,596,000	27,197,000	7,890,000	43,425,000	111,331,000
% total acres	4%	26%	24%	7%	39%	100%

South Central Forest Owners by Type of Entity

Ownership Type	Ownerships	% Ownership	Acreage	% Acreage	Avg. Acreage
Forest industry	2,100	0%	5,443,000	6%	2,592
Farm	728,300	29%	32,322,000	35%	44
Industrial business	2,900	0%	3,095,000	3%	1,067
Real estate	30,400	1%	4,371,000	5%	144
Nonindustrial business	7,000	0%	713,000	1%	102
Recreation club	26,300	1%	1,721,000	2%	65
Public utility*	25	0%	288,000	0%	11,520
Individuals	1,648,400	66%	39,502,000	42%	24
Other	53,900	2%	5,879,000	6%	109
Total	**2,499,325**	**100%**	**93,334,000**	**100%**	**37.34**

*Estimated ownerships with an error of plus or minus 25

Tables B-9. (*continued*)

Primary Reason for Owning Forestland for the South Central

Reason	Ownerships	% Owners	Acres	% Acres
Timber	93,400	4%	38,079	37%
Land investment	229,100	11%	13,310	13%
Farm and domestic uses	264,200	12%	8,368	8%
Part of residence	656,600	31%	9,307	9%
Part of farm	305,800	14%	9,901	10%
Recreation	165,600	8%	8,371	8%
Enjoyment	202,000	9%	6,284	6%
Other	73,400	3%	6,274	6%
No answer	151,900	7%	1,861	2%
Total	2,142,000	100%	101,755	100%

Figures do not include estates

South Central Forestland Ownership by Date of Acquisition

Years	Total Owners	% Owners	Acres	% Acres
1990–1994	325,300	13%	6,557,000	6%
1980–1989	626,900	25%	16,963,000	15%
1970–1979	555,200	22%	18,178,000	16%
1960–1969	273,300	11%	18,211,000	16%
1950–1959	180,000	7%	12,795,000	11%
1940–1949	103,800	4%	9,307,000	8%
1901–1939	111,100	4%	11,995,000	11%
Before 1900	4,500	0%	8,416,000	8%
No answer	319,100	13%	8,909,000	8%
Total	2,499,200	100%	111,331,000	100%

South Central Forestland Ownership by Occupation

Occupation	Total Owners	% Owners	Acres	% Acres
White collar	651,800	30%	20,835,000	20%
Blue collar	373,000	17%	5,467,000	5%
Farmer	207,600	9%	10,181,000	10%
Service worker	80,000	4%	1,087,000	1%
Homemaker	21,900	1%	1,943,000	2%
Retired	757,500	34%	23,913,000	23%
Other	117,300	5%	42,354,000	40%
Total	**2,209,100**	**100%**	**105,780,000**	**100%**

Figures do not include landowners who did not respond

Ownership Expectations of Future Timber Harvest for South Central

Expected Future Harvest	% Owners	% Acres
1–10 years	30%	60%
Indefinite	29%	26%
Never	35%	12%
No answer	6%	2%

Table B-10

State Government Programs Focused on Major Private Forestry Objectives by Activity, Region, and Type of Program (1992)

Major forestry activity and type of program	NE	LS	MA	MC	SE	SC	GP	RM	West	Total
Protect										
Water Quality										
Educational programs	6	3	6	5	5	5	5	5	6	46
Technical assistance	6	3	7	5	5	5	5	6	5	47
Voluntary guidelines	5	3	6	4	5	5	1	4	1	34
Tax incentives	1	1	4	3	0	1	3	1	0	14
Fiscal incentives	2	3	5	3	1	4	5	4	2	29
Regulatory programs	6	1	5	1	4	1	0	2	6	26
Promote										
Reforestation										
Educational programs	6	3	6	5	6	5	4	5	6	46
Technical assistance	6	3	6	5	6	5	5	6	4	46
Voluntary guidelines	1	1	3	2	3	3	3	4	1	15
Tax incentives	2	3	3	3	1	1	0	1	2	16
Fiscal incentives	5	2	5	3	4	5	5	5	3	39
Regulatory programs	3	0	4	0	0	0	0	1	6	14
Improve Timber										
Harvesting Methods										
Educational programs	6	3	6	5	5	4	5	5	6	45
Technical assistance	6	3	7	5	6	5	5	6	4	47
Voluntary guidelines	4	2	6	1	3	3	2	4	2	27
Tax incentives	2	2	3	1	0	1	0	0	0	09
Fiscal incentives	3	0	4	0	0	1	2	2	1	13
Regulatory programs	4	0	4	0	1	1	0	0	6	16

Tables B-9. (*continued*)

	NE	LS	MA	MC	SE	SC	GP	RM	West	Total
Protect from Wildfire, Insects, and Diseases										
Educational programs	6	3	6	5	5	5	5	6	6	47
Technical assistance	6	3	7	4	6	5	4	6	6	46
Voluntary guidelines	3	0	3	1	2	3	2	4	2	20
Tax incentives	0	1	3	2	0	0	0	0	0	06
Fiscal incentives	1	1	4	2	1	0	2	4	2	17
Regulatory programs	5	2	3	1	3	2	1	4	6	27
Protect Wildlife and Rare and Endangered Species										
Educational programs	6	3	7	5	6	5	4	5	5	46
Technical assistance	5	3	6	5	6	5	5	5	4	45
Voluntary guidelines	4	1	3	1	1	2	2	2	2	16
Tax incentives	0	0	1	2	0	0	0	0	0	03
Fiscal incentives	3	2	5	3	2	4	5	2	2	26
Regulatory programs	4	2	2	0	3	1	1	2	5	20
Enhance Recreation and Aesthetic Qualities										
Educational programs	6	3	6	4	5	5	4	5	3	42
Technical assistance	6	3	7	5	5	5	5	6	3	45
Voluntary guidelines	3	1	2	1	1	2	2	2	2	16
Tax incentives	1	1	1	2	0	1	0	1	1	06
Fiscal incentives	4	1	6	2	2	4	2	3	1	25
Regulatory programs	2	0	1	0	0	0	0	0	0	06

Northeast: CT, ME, MA, NH, RI, VT; **Lake States:** MI, MN, WI; **Mid-Atlantic:** DE, D, NJ, NY, PA, VA, WV; **Mid-Continent:** IL, IN, KY, MO, OH; **Southeast:** AL, FL, GA, MS, NC, SC; **South Central:** AR, LO, OK, TN, TX; **Great Plains:** IA, KS, ND, SD; **Rocky Mountain:** AZ, CO, MT, NM, UT, WY **West:** AK, CA, HI, ID, NV, OR, WA
Source: Ellefson 1995. Regulation of Private Forestry Practices by State Government.

Appendix C.
Contact List of Organizations Referenced

Adirondack Land Trust–The Nature Conservancy
Contact: *Michael Carr*–Director
PO Box 65
Keene Valley, NY 12943
Ph: (518) 576-2082
dfeeley@tnc.org
http://www.tnc.org/adirondacks/

Adopt-a-Watershed
Contact: *Kim Stukley*–Executive Director
PO Box 1850 Hayfork, CA 96041
Ph: (530) 628-5334
Fax: (530) 628-4212
kim@adopt-a-watershed.org
http://www.adopt-a-watershed.org/

American Forest & Paper Association–Sustainable Forestry Initiative
Contact: *Rick Cantrell*–Director
1111 Nineteenth Street, NW, Suite 800
Washington, DC 20036
Ph: (202) 463-2432
rick_cantrell@afandpa.org
http://www.afandpa.org/forestry/sfi_frame-html

The American Tree Farm System
Shared Streams and Forest Flyways
Contact: *Drue DeBerry*
Ph: (888) 889-4466
DrueDeBerry@affoundation.org
http:www.treefarmsystem.org

227

Applegate Partnership
Contact: *Jack Shipley*
CPO Box 3277
Applegate, OR 97530

Association of Consulting Foresters, Inc.
Contact: *Lynn Wilson*–Administrative Director
732 N. Washington Street, Suite 4-A
Alexandria, VA 22314
Ph: (703) 548-0990
Fax: (703) 548-6395
director@acf-foresters.com
http://www.acf-foresters.com/

The Conservation Fund
1800 N. Kent Street, Suite 1120
Arlington, VA 23209-2156
Ph: (703) 525-6300
Fax: (703) 525-4610
http://www.conservationfund.org/conservation/index.html

Ducks Unlimited
Contact: *Julius F. Wall*–President
One Waterfowl Way
Memphis, TN 38120
Ph: (800) 453-8257
conserv@ducks.org
http://www.ducks.org/

Ecotrust
Contact: *Ed Backus*–Director of Community Projects
1200 NW Naito Parkway, Suite 47
Portland, OR 97209
Ph: (503) 227-6225
Fax: (503) 222-1517
info@ecotrust.org
http://www.ecotrust.org/

The Forest Bank–The Nature Conservancy
Contact: *Kent Gilges*–Director
339 East Avenue, Suite 300
Rochester, NY 14604
Ph: (716) 232-3530
Fax: (716) 546-7825
kgilges@tnc.org

Forest Society of Maine
Contact: *Alan Hutchinson*–Executive Director
PO Box 775
Bangor, ME 04402
Ph: (207) 945-9200
Fax: (207) 945-9229
alanhfsm@mint.net
http://mltn.org/trusts/fsm.htm

Forest Stewards Guild–Forest Trust
Contact: *Steve Harrington*–Coordinator
PO Box 519
Sante Fe, NM 87504
Ph: (505) 983-8992
Fax: (505) 986-0798
info@foreststewardsguild.org
http://www.foreststewardsguild.org/

Forest Stewardship Council
134 Twenty-Ninth Street NW
Washington, DC 20037
Ph: (202) 392-0413
Toll Free: (887) 372-5646
Fax: (202) 342-6589
http://wwwfscus.org/

The Game of Logging®
Contact: *Tim Ard*
PO Box 1048
Hiram, GA 30141-1048
Ph: (770) 943-4745
Info@forestapps.com

Hiawatha Sustainable Woods Cooperative
PO Box 248
Fountain City, WI 54629
Ph: (608) 687-8193

Inforain–Ecotrust
Contact: *Mike Mertens*–GIS Analyst
1200 NW Naito Parkway, Suite 47
Portland, OR 97209
Ph: (503) 227-6225
Fax: (503) 222-1517
info@ecotrust.org

Institute for Agriculture and Trade Policy (IATP)–Community Forestry Resource Center
Contact: *Phillip Guillery*–Forest Program Director
2105 First Avenue South
Minneapolis, MN 55404
Ph: (612) 870-3456
pguillery@iatp.org
http://www.forestrycenter.org

Institute for Sustainable Forestry
PO Box 1580
Redway, CA 95560
Ph: (707) 247-1101
Fax: (707) 247-3555
info@isf-sw.org
http://www.isf-sw.org/aboutISF.htm

Land Trust Alliance
Contact: *Jean Hocker*–President
1319 F Street NW, Suite 501
Washington, DC 20004
Ph: (202) 638-4725
Fax: (202) 638-4730
jhocker@lta.org
http://www.lta.org/

Logger Education to Advance Professionalism (LEAP)
Contact: *George Miller*–Safety Director
PO Box 67
Coeur d'Alene, ID 83816
Ph: (208) 667-6473
Fax: (208) 667-2144
alclog@aol.com

Longleaf Alliance
Contact: *Dean Gjerstad*
School of Forestry
Auburn University, AL 36849
Ph: (334) 844-1020
Fax: (334) 844-1084
gjerstad@forestry.auburn.edu

Minnesota Forest Resources Council
2003 Upper Buford Circle
St. Paul, MN 55108
Ph: (651) 603-0109
Fax: (651) 603-0110
http://www.frc.state.mn.us

Monadnock Landscape Partnership
Contact: *Benjamin Mahnke*–Land Protection Specialist
PO Box 337
Keene, NH 03431-0337
Ph: (603) 357-0600
conservancy@monad.net

Mountain Association for Community Economic Development (MACED)
Contact: *Mike Jenkins*–Director of Sustainable Forestry
433 Chestnut Street
Berea, KY 40403
Ph: (606) 986-2373, Ext. 214
Fax: (606) 986-1299
mjenkins@maced.org
http://www.maced.org/

National Wildlife Federation
8925 Leesburg Pike
Vienna, VA 22184
Ph: (703) 790-4000
http://www.nwf.org/

National Woodland Owners Association
Contact: *Keith Argow*–President
374 Maple Avenue E, Suite 210
Vienna, VA 22180
Ph: (703) 255-2700
Fax: (703) 281-9200
nwoa@mindspring.com
http://www.wood-land.org

The Nature Conservancy
4245 North Fairfax Drive, Suite 1000
Arlington, VA 22203-1606
Ph: (703) 841-5300
http://www.tnc.org/

New England Forestry Foundation
Contact: *Keith Ross*–Vice President
238 Old Dunstable Road
Groton, MA 01450
Ph: (978) 448-8380, Ext. 109
Fax: (978) 448-8379
kross@neforestry.org

Northern Forest Alliance
Andria Colnes–Director
43 State Street
Montpelier, VT 05602
Ph: (802) 223-5256
Fax: (802) 229-4642
acolnes@nfainfo.org

Oregon Forest Resource Trust
Contact: *Jim Cathcart*–Forest Resource Trust Manager
2600 State Street
Salem, OR 97310
Ph: (503) 945-7380
Fax: (503) 945-7376
jcathcart@odf.state.or.us
http://www.odf.state.or.us/fa/SF/FRT/FRT.htm

The Pacific Forest Trust
Contact: *Laurie Wayburn*–President
PO Box 879
Boonville, CA 95415
Ph: (707) 895-2090
Fax: (707) 895-2138
lwayburn@pacificforest.org

Rainkist
Contact: *Arlen Veleke*
3515 College Street NE
Lacey, WA 98503
Ph: (877) 284-4027
Fax: (360) 413-9450
http://rainkist.org/

Ruffed Grouse Society
451 McCormick Road
Coraopolis, PA 15108
Ph: (412) 262-4044
Fax: (412) 262-9207

Sandhills Area Land Trust
Contact: *Nell Allen*–Administrator
PO Box 1032
Southern Pines, NC 28388-1032
Ph: (910) 695-3322
http://metalab.unc.edu/ctnc/trusts/sandhills.html

Save-the-Redwoods League
Contact: *Kate Anderton*
114 Sansome Street, Room 605
San Francisco, CA 94104-3814
Ph: (415) 362-2352
Fax: (415) 362-7017
http://www.savetheredwoods.org/

See-the-Forest
Contact: *Shauna Ratner*–Principal of YellowWood Associates
95 South Main Street
St. Albans, VT 05478
Ph: (802) 524-6141
Fax: (802) 524-6643
yellow2@together.net
http://www.together.net/~yellow

Shorebank Pacific Enterprises–Shorebank Pacific
Contact: *Mike Dickerson*–Deputy Director
PO Box 826
Ilwaco, WA 98624
Ph: (360) 642-4265
Fax: (360) 642-4078
cab@sbpac.com
http://www.sbpac.com/

Sierra Business Council
PO Box 2428
Truckee, CA 96160
Ph: (530) 582-4800
Fax: (530) 582-1230
sbcinfo@sbcouncil.org
http://www.tahoe.ceres.ca.gov/sbc/

The Society for Protection of New Hampshire Forests
54 Portsmouth Street
Concord, NH 0331-5400
Ph: (603) 224-9945
Fax: (603) 228-0423
skh@spnhf.org
http://www.spnhf.org/

Society of American Foresters
Contact: *Bill Banzhaf*–Executive Vice President
5400 Grosvenor Lane
Bethesda, MD 20814-2198
Ph: (301) 897-8720, Ext. 120
Fax: (301) 897-3690
www.safnet.org

Sonoma County Open Space and Agriculture Protection
747 Mendocino Avenue, Suite 100
Santa Rosa, CA 95401
Ph: (707) 524-7360
Fax: (707) 524-7370
http://www.sonoma-county.org/services

Sustainable Forestry Partnership
Contact: *Richard A. Fletcher*–Associate Director
Oregon State University
154 Peavy Hall
Corvallis, OR 97331
Ph: (541) 737-4991
Toll Free: (877) SFP-4937
Fax: (541) 737-3385
sfp@cof.orst.edu
http://sfp.cas.edu/

Sustainable Woods Cooperative
Contact: *Jim Birkemeier*–Timber Green Forestry
511478 Soeblner Road
Spring Green, WI 53588
Ph: (608) 588-7342
Fax: (608) 588-7651
jim@timbergreenforestry.com
http://www.iatp.org/forestry/

Tall Timbers Research Station–Red Hills Conservancy
Contact: *Kevin McGorty*–Red Hills Program Director
13093 Henry Beadel Drive
Tallahassee, FL 32312-0918
Ph: (805) 893-4153, Ext. 238
Fax: (805) 668-7781
kmcgorty@ttrs.org

The Trust for Public Land
116 New Montgomery Street, 4th Floor
San Francisco, CA 94105
Ph: (415) 495-4014
Fax: (415) 495-4103
http://www.tpl.org/

Vermont Coverts Inc.
421 Governor Chittenden Road
Williston, VT 05495
Ph: (805) 878-2180

Vermont Family Forests
Contact: *David Brynn*
1590 Route 7 Street
Middlebury, VT 05753
Ph: (802) 388-4969

The Vermont Land Trust
Contact: *Gil Livingston*–Vice President of Land Conservation
8 Bailey Avenue
Montplier, VT 05602-2101
Ph: (802) 223-5234
Fax: (802) 223-4223
gil@vlt.org

Water Watch
Contact: *Reed Benson*
213 Southwest Ash, Suite 208
Portland, OR 97204
Ph: (503) 295-4039
Fax: (503) 295-2791
reed@waterwatch.org
http://www.waterwatch.org/

Wisconsin Family Forests
Contact: *Martin Pionke*–Chair
PO Box 99
Hancock, WI 54943
Ph: (715) 249-5406

Appendix D.
Examples of Recent Nonprofit Forest Conservation Transactions

1. Champion Forestlands Transaction
"Largest Conservation Partnership in U.S. History Completes Purchase of 300,000 Acres" —from *ForestWeb News*

Stamford, Conn.—Aug. 6, 1999—Completing the largest multi-state conservation project in U.S. history, Champion International Corporation and The Conservation Fund today announced the sale of 106,100 acres of land in Vermont for $20 million, capping almost 300,000 acres of acquisitions that began last month in New York and New Hampshire. The project was first announced by the governors of New York and Vermont in December, 1998.

In today's transaction, approximately 84,000 acres of timberland in Vermont, subject to conservation and public access easements, went to Essex Timber Company, LLC for $7.5 million. Through a gift from the Richard King Mellon Foundation of $4.5 million, Vermont's Agency of Natural Resources received an additional 16,500 acres. With $4.5 million appropriated by the legislature, the State of Vermont, aided by the Freeman Foundation, other foundations, and individuals, secured public access and conservation easements covering the 84,000 acres of forestry land. The Conservation Fund will temporarily hold 5,600 acres slated for the state of Vermont, dependent on a future federal grant requested under the North American Wetlands Conservation Act.

To complete this complex transaction, The Conservation Fund formed partnerships with private investors, foundations, public agencies, and the nonprofit conservation community, the Vermont Land Trust being the lead partner, to secure funding and to provide long-term management.

"We've secured more than $70 million in private and public funds

236

since December," said Patrick F. Noonan, chairman of The Conservation Fund. "It has taken teamwork at every level. In Vermont alone, The Conservation Fund and the Vermont Land Trust have identified $28 million in public and private funds to purchase the 133,000 acres of land from Champion."

Seward Prosser Mellon, president of the Richard King Mellon Foundation, said, "The Foundation's American Land Conservation Program, under which this acquisition is being made, reflects our Foundation's— and family's—traditional and continuing interest in land conservation. In these times of tight budgets, we feel that the private sector has an opportunity and an obligation to augment the conservation work of state and federal agencies. We are pleased that our program will help safeguard wildlife habitat, forests, and wetlands in Vermont."

Established in 1947, the Richard King Mellon Foundation, based in Pittsburgh, Pennsylvania, has a long-standing commitment to conservation. Since 1977 the Foundation has made major grants and gifts for conservation primarily in the areas of land acquisition, wetlands protection, and wildlife preservation.

On July 1, 1999, New York Governor George E. Pataki announced the purchase of 29,000 acres of environmentally sensitive land to be added to the Andirondack Forest Preserve and of an easement on the surrounding 110,000 acres that insures public access and environmentally responsible timber management. The state paid $24.9 million for the land and easement. Simultaneously, The Forestland Group, a firm specializing in timber investments, acquired the 100,000 acres of commercial timberland for long-term, sustainable forestry operations for approximately $21 million.

On July 16, 1999, The Conservation Fund bought Champion's 18,600 acres in New Hampshire, for $3.75 million, which includes timber and high elevation land.

On July 23, 1999, the U.S. Fish and Wildlife Service purchased 26,000 acres in northern Vermont's environmentally significant Nulhegan Basin on the Connecticut River for $6.5 million.

"Champion Forestlands Have New Owners" Vermont Land Trust (July 1999, www.vlt.org/CInewpress.html)

The Conservation Fund, a national land and water conservation organization based in Virginia, and the Vermont Land Trust (VLT) have completed the purchase of 133,000 acres from Champion International, Inc. in northeastern Vermont. At a simultaneous closing, The Fund took title to 106,100 acres and reconveyed, subject to easements, 100,500 acres to a private timberland investor and the Vermont Agency of Natural Resources. The announcement followed the July 21st purchase of 26,000 acres by the United States Fish and Wildlife Service (USFWS).

The sale by Champion and the transfer to new owners represents a new future for a property which has been integral to the Northeast Kingdom's economy and heritage. Essex Timber Company, LLC paid $7.5 million for approximately 84,000 acres of timber land. The land is restricted by conservation and public access easements, which will ensure sustainable harvesting practices, continued public. recreational access, and protection of the unique biological attributes associated with the 84,000 acres and the adjoining public lands. The agreement also provides for continued leasing of privately owned camps. Essex Timber Company (ETC) is based in Boston, Massachusetts. North Country Environmental and Forestry of Concord, Vermont has contracted to be ETC's forest manager.

Wilhelm Merck, managing member of ETC, said, "This is an exceptional opportunity for my investors, who are serious about growing high quality sawtimber for long-term gains. Thanks to the presence of the easements, our objectives can be matched with the capacity of the forest resource, and the public's wishes for the property are well served also."

The heritage of this property and the region's economy has been one of "boom and bust," tracking cycles of massive harvests followed by reduced activity as the forest renewed itself. The conservation easement requires management that results in a sustained yield of high quality sawtimber, a dramatic change from the fiber-based goals of the papermills in the past, and a change that bodes well for the future of the region.

The Vermont Land Trust will have the primary responsibility of administering the working forest easement on the private lands. Darby Bradley, VLT's president, has known Wil Merck through his involvement with the New England Forestry Foundation. "I am very pleased to be able to work with Wil and the Essex Timber Company," said Bradley. "When we put the 84,000 acres on the market, we wanted to find more than just an investor—somebody who would see the potential that these lands have to contribute to the economy and communities of the region. We did that."

The sale marks the near completion of two years of intensive effort involving many non-profit and governmental organizations. In addition to The Fund and the Vermont Land Trust, The Nature Conservancy of Vermont (TNC), Agency of Natural Resources (ANR), Vermont Housing and Conservation Board (VHCB), and US Fish and Wildlife Service all played critical roles in the negotiations. Governor Howard Dean and Vermont's congressional delegation were also instrumental in helping secure portions of the funding necessary for the project.

The Fund and VLT were able to complete the $26.5 million purchase by pulling together a wide range of funding sources. In addition to the purchases by ETC and USFWS, the Freeman Foundation made a $4 million grant, the State of Vermont appropriated $4.5 million, and the Richard King Mellon Foundation made a matching grant of $4.5 million. Other foundations and individuals made contributions to cover a portion of the transaction costs.

In addition to the sale of 84,000 acres to the Essex Timber Company, the transaction also included the transfer of approximately 16,500 acres to the State of Vermont. The Fund hopes to transfer the final 5,600 acres to the Agency of Natural Resources if the last piece of the funding package becomes available in the fall. The State lands will be managed by and are subject to a conservation easement co-held by The Nature Conservancy of Vermont and the Vermont Housing and Conservation Board. TNC led a process that identified the most ecologically significant portions of the Champion lands. The Fund and VLT used the results to decide which land should be placed in public ownership, either by the State of Vermont or USFWS. USFWS purchased 26,000 acres in the Mulhegan Basin on July 21st for inclusion in the Silvio O. Conte National Wildlife Refuge. The State and federal lands include many rare species, the largest deer-yard in the state, several exemplary wetland natural communities, a number of remote ponds, and many miles of riparian habitat.

In addition to the array of natural resources on the publicly held property, access to the public lands will complement the recreational resources located on the private working forest lands. ANR, in collaboration with TNC, will engage in a thorough public process as access plans and management plans are developed, which will consider both the ecological values and habitat needs of the parcel.

When the Vermont Legislature appropriated $4.5 million for the Champion project, the Vermont Housing and Conservation Board was charged with looking out for the State's interests in the negotiations.

The former Champion lands comprise 31% of Essex County. They are located in Averill, Avery's Gore, Bloomfield, Brighton, Brunswick, East

Haven, Ferdinand, Granby, Lemington, Lewis, Maidstone, and Victory as well as in the towns of Morgan in Orleans County and Burke in Caledonia County. The sale of this property represents the last major holding of a paper company in Vermont, reflecting an industry trend to divest in the Northeast and invest in the southern United States and South America.

2. Atlas Forestlands Transaction

"The Nature Conservancy of Vermont and the Vermont Land Trust Acquire the Third Largest Private Forest Holding in the State"
—Vermont Land Trust (1999, www.tnc.org)

The Nature Conservancy of Vermont and the Vermont Land Trust recently completed one of the largest private conservation projects ever in the eastern United States. Both organizations are pleased to announce the purchase of the third largest private forest holding in Vermont from the Atlas Timberland Company for $5,572,000 ($208/acre). On December 29, 1997 the partnership, known as the Atlas Timberlands Partnership (ATP), acquired 26,789 acres of undeveloped working forest land located primarily in Vermont's northern Green Mountains.

Funding for the purchase came entirely from private sources. The organizations received a $5,000,000 grant from the Freeman Foundation and a $572,000 loan from the Conservancy's Land Acquisition Fund. The Partners are launching a fundraising campaign to repay the loan.

The acquisition includes 23 parcels, ranging in size from 10 to 2,636 acres, spread across 16 towns. The parcels are located in the towns of Bakersfield, Belvidere, Craftsbury, Eden, Elmore, Groton, Hardwick, Jay, Lowell, Montgomery, Orange, Plainfield, Richford, Westfield, Westmore, and Wolcott in Vermont, as well as one in Alton, New York. More than half of the holdings are concentrated in two parcel groups—12,500 acres in Bakersfield, Belvidere, Eden, Lowell, and Montgomery, and 4,421 acres in Jay and Richford. Wagner Forest Management, LLC of Lyme, NH and Newport, VT managed the land for Atlas Timberland Company.

The Partnership ensures the Atlas lands will remain intact and part of Vermont's working forest. ATP will continue to both carefully manage the lands for a stable supply of high quality timber and protect the biological integrity of the forest. The two groups believe timber production can be compatible with environmental protection, scenic quality, and recreation. ATP will retain Wagner through an interim contract to manage timber harvesting this winter (1997–98). Traditional public use of the land for hunting, fishing, hiking, and snowmobiling will continue and the Partnership will pay all local and State property taxes.

Bob Klein, Director of the Vermont chapter of The Nature Conservancy (TNC), explains how the Partnership combines key aspects of each organization's mission. "This project protects those portions of the forested landscape where the missions of the Vermont Land Trust and TNC overlap. TNC focused historically on the protection of biological resources without a great deal of thought to timber production, while VLT emphasized the economic use of the forest. We realize these are not separate directions if we are to maintain the integrity of Vermont's ecosystems and forests in the future. Vermont's history showed a balance between ecology and economics long before the term sustainability was in vogue. ATP will help us understand how to maintain this balance in the future, so that the natural world Vermonters enjoy will thrive."

ATP's commitment to actively manage its timber holdings, protect and enhance biological resources, and maintain traditional public uses establishes a new precedent and model for future land agreements. ATP may trade standing timber or conserved forest land for conservation easements on other properties that mix important biological and forest resources. This innovative approach will test the practicality of using timber as a "renewable conservation currency."

Darby Bradley, President of Vermont Land Trust and Chair of Vermont's Forest Resources Advisory Council notes that the forest products industry is critical to the economy of rural Vermont. Referring to FRAC's final report issued last September, Bradley cites the fact that the forest industry is Vermont's largest employer in the manufacturing sector.

"This industry provides over 8,000 jobs to Vermonters, including 6,000 jobs in furniture-making and other finished wood products," says Bradley. "With the ownership of so much of Vermont's woodlands in a state of flux, we must find ways to maintain large blocks of woodland to provide the raw materials for the industry, protect our environment, and preserve a way of life for Vermonters."

The Partners pursued this opportunity for a variety of reasons including parcel size, location and the quality of management performed by

Wagner Forest Management, LTD. Carl Szych, a Newport resident and WFM vice-president, has been involved with the Atlas lands for the past eighteen years. In his words, "The Atlas properties should remain forest land to provide timber products and jobs while protecting the environment for future generations. This purchase goes a long way to ensuring that future."

In the weeks ahead, ATP project managers plan to meet with local communities, forest products industry representatives, public officials, environmental groups, and others to explain their plans and listen to ideas and concerns. ATP will evaluate the timber resources, ecology, recreational uses, and other qualities of each parcel as it develops future management plans for the properties.

3. Upper St. John River (Maine) Forestlands Transaction "Nature Conservancy Buys 185,000 Acres of Maine Forest Protects 40 Miles of Upper St. John River from Development" —(Jan. 5, 1999, www.tnc.org)

Augusta, Maine—The Nature Conservancy today announced the $35.1 million purchase of 185,000 acres of remote Maine forest. This transaction represents the largest single conservation acquisition in the state's history and the largest tract of land ever purchased for conservation in the northeastern United States.

The lands, in the far northwestern corner of the state, encompass some 286 square miles of unbroken forest, including a 40-mile stretch of the Upper St. John River, the longest free-flowing river east of the Mississippi.

"The Upper St. John is one of the most admired wilderness rivers in the country," said Ken Wommack, Executive Director of The Nature Conservancy of Maine. "These forests, rivers and wetlands teem with wildlife and are rich in remote recreational opportunities. We intend to keep this resource undeveloped, undiminished and open forever for the people of Maine."

The Conservancy is purchasing these lands from International Paper on the open market. It will close on the property before year's end and, during the coming months, will develop a comprehensive management plan for the tract. As a first step, the Conservancy has retained Wagner Woodlands, one of the most respected forest land managers in the Northeast, to help manage the property.

These measures will support the organization's long-term vision for the Upper St. John watershed, which includes a mix of protected lands, especially along the river corridor, with compatibly managed forest lands. All of the Conservancy's property will be open for traditional uses, such as hunting, fishing and recreation.

To develop long-term protection strategies for the land, the Conservancy will work in partnership with Maine Coast Heritage Trust, a statewide land trust, and other conservation partners, including timber companies in the region.

The lands encompass a third of the Upper St. John River, a 120-mile, free-flowing wilderness river that flows north from Baker Lake to the confluence with the Allagash River. The watershed is known for its abundance of moose, bear and other wildlife and boasts the second highest concentration of rare plants in Maine. It is home to the Furbish lousewort, a globally endangered plant found nowhere else on earth. Except for a handful of primitive lean-tos, this area shows few signs of civilization.

The newly acquired lands fall within a large natural region, or ecoregion, the Conservancy calls the Northern Appalachian/Boreal Forest, which sweeps down out of Canada, blankets more of Maine, and flows over northern New Hampshire and Vermont and on to the Adirondacks of New York.

References

AF&PA. 1999. *SFI 1999 4th Annual Progress Report on the American Forest and Paper Association's Sustainable Forestry Initiative Program*. Washington, DC: American Forest and Paper Association.

Alig, Ralph J. 1986. Econometric Analysis of the Factors Influencing Forest Acreage Trends in the Southeast. *Forest Science* 32(1): 119–134.

————. 2000. *Draft 2000 RPA Assessment: Summary of Findings From Area Change Analyses and Projections*. Corvallis, OR: USDA Forest Service.

Alig, Ralph J., and R. G. Healy. 1987. Urban and Built-up Land Area Changes in the United States: An Empirical Investigation of Determinants. *Land Economics* 63(3): 215–226.

Alig, Ralph J., and David N. Wear. 1992. *Changes in Private Timberland: Statistics and Projections for 1952 to 2040. Journal of Forestry* 90(5): 31–36.

Alig, Ralph J., Karen J. Lee, and Robert J. Moulton. 1990a. *Likelihood of Timber Management on Nonindustrial Private Forests: Evidence from Research Studies*. General Technical Report SE-60. Asheville, NC: U.S. Department of Agriculture, USDA Forest Service, Southeast Forest Experiment Station.

Alig, Ralph J., William G. Hohenstein, Brian C. Murray, and Robert G. Haight. 1990b. *Changes in Area of Timberland in the United States, 1952–2040, by Ownership, Forest Type, Region, and State*. General Technical Report SE-64. Asheville, NC: U.S. Department of Agriculture, USDA Forest Service, Southeast Forest Experiment Station.

Alig, Ralph J., Michael T. Dicks, and Robert J. Moulton. 1999. *Land Use Dynamics Involving Forestland: Trends in the U.S. South*. Presented at the Southern Forest Economics Workshop. Williamsburg, VA, March 20–23, 1998.

American Farmland Trust. 1993. Open Space and Taxes (from a Land Conservation Coalition of Connecticut) Summary: Does Conservation Pay? Symposium of the Lincoln Institute of Land Policy, May 12, 1992. In *Connecticut Forest & Park Association Newsletter. Connecticut Woodlands,* Spring 1993.

Araman, P. A., and J. B. Tansey. 1990. The US Hardwood Situation Related to Exports. Pages 41–51 in *Proceedings of the 1990 Joint International Conference on Processing and Utilization of Low-Grade Hardwoods and International Trade of For-*

est-Related Products. Sponsored by the National Taiwan University, Graduate Institute of Forestry, Republic of China, and Auburn University, School of Forestry, Auburn, AL.

Argow, Keith A. 1999. President, National Woodland Owners Association. Telephone interview, July 13, 1999.

Barbour, Michael G., and William D. Billings. 1988. *North American Terrestrial Vegetation.* New York: Cambridge University Press.

Baughman, Melvin J. 1993. *Effectiveness of Forestry Incentive Program.* Presentation at NA Forest Resource Program Leaders/Extension Foresters Meeting. Amana, IA, June 10, 1993.

Best, Constance, and Michael Jenkins. 1999. *Opportunities for Investment: Capital Markets and Sustainable Forestry.* Boonville, CA: The Pacific Forest Trust; Chicago, IL: John D. and Catherine T. MacArthur Foundation.

Binkley, Clark S. 1981. *Timber Supply from Private Nonindustrial Forests: A Microeconomics Analysis of Landowner Behavior.* School of Forestry and Environmental Studies Bulletin No. 92. New Haven, CT: Yale University, School of Forestry and Environmental Studies.

Birch, Thomas W. 1996. *Private Forest-land Owners of the United States, 1994.* Resource Bulletin NE-134. Radnor, PA: U.S. Department of Agriculture, USDA Forest Service, Northeastern Forest and Experiment Station.

Birdsey, Richard, A., and Linda S. Heath. 1995. *Productivity of America's Forests and Climate Change.* Gen. Tech. Rep. RM-271 Ft. Collins, CO: U.S. Department of Argiculture, Forest Service, Rocky Mountain Research Station.

Bliss, John C., Sunil K. Nepal, Robert T. Brooks Jr., and Max D. Larsen. 1994. *Forestry Community or Granfallon?: Do Forest Owners Share the Public's Views? Journal of Forestry* 92(9): 6–10.

———. 1997. In the Mainstream: Environmental Attitudes of Mid-South Forest Owners. *Southern Journal of Applied Forestry* 21(1): 1–7.

Bonnie, R. 1997. Safe Harbor for the Red-Cockaded Woodpecker. *Journal of Forestry* 95(4): 17–22.

Boulinier, T., J. D. Nichols, J. E. Hines, J. R. Sauer, C. H. Flather, and K. H. Pollock. 1998. Higher Temporal Variability of Forest Breeding Bird Communities in Fragmented Landscapes. In *Proceedings of the National Academy of Sciences* 95: 7497–7501.

Bourke, Lisa, and Albert E. Luloff. 1994. Attitudes Toward the Management of Nonindustrial Private Forest Land. *Society and Natural Resources* 7: 445–457.

Boyd, R. 1983. The Effects of FIP and Forestry Assistance on Nonindustrial Private Forests. In: J. P. Royer and C. D. Risbrudt (eds.), *Proceedings of Conference on Nonindustrial Forests: A Review of Economic and Policy Studies.* Durham, NC: Duke University.

———. 1984. Government Support of Nonindustrial Production: The Case of Private Forests. *Southern Economics Journal* 51(July 1984): 89–107.

Brown, Thomas C., and Dan Binkley. 1994. *Effect of Management on Water Quality in North American Forests.* General Technical Report RM-248. Fort Collins, CO: U.S. Department of Agriculture, USDA Forest Service, Rocky Mountain Forest and Range Experiment Station.

Brunson, Mark W., Linda E. Kruger, Catherine B. Tyler, and Susan A. Schroeder. 1996. Defining Social Acceptability in Ecosystem Management. General Technical Report PNW-GTR-369. Proceedings at a Workshop held in Kelso WA, June 23–25, 1992. Portland, OR: U.S. Department of Agriculture, USDA Forest Service, Pacific Northwest Research Station.

Burgess, R. L., and D. M. Sharpe, eds. 1981. *Forest Island Dynamics in Man-dominated Landscapes.* New York: Springer-Verlag.

Campbell, G. E. 1988. *The Impacts of Cost-Sharing and Income Taxation on Forestry Investments: A Case Study.* FSR 88-11. Urbana: University of Illinois Agricultural Experiment Station.

Carey, Andrew B., and Catherine Elliot. 1994. *Washington Forest Landscape Management Project-Progress Report.* Olympia, WA: Washington State Department of Natural Resources.

Chang, S. Joseph. 1996. US Forest Property Taxation Systems and Their Effects. In *Proceedings: Symposium on Non-Industrial Private Forests: Learning from the Past, Prospects for the Future.* Washington, DC, February 18–20, 1996.

Citizens For New Hampshire Land & Community Heritage. 1999. *New Hampshire's Land and Community Heritage at Risk, February 1999.* Concord, NH: Citizens For New Hampshire Land & Community Heritage.

Collins, Charles H. 1999. Remarks by Charles H. Collins, the Forestland Group LLC to the Empire State Forest Products Association, October 6, 1999. Cambridge, MA: The Forestland Group LLC.

Constanza, R., R. d'Arge, R. de Groot, S. Farber, M. Grasso, B. Hannon, K. Limberg, S. Naeem, R. V. O'Neill, J. Paruelo, R. G. Raskin, P. Sutton, and M. van den Belt. 1997. The Value of the World's Ecosystem Services and Natural Capital. *Nature* 387: 253–260.

Crossely, Rachel, and Johnathan Points. 1998. *Investing in Tomorrow's Forests: Profitability and Sustainability in the Forest Products Industry.* Washington, DC: World Wildlife Fund.

Cubbage, Frederick W., Thomas G. Harris Jr., David N. Wear, Robert C. Abt, and Gerardo Pacheco. 1995. Timber Supply in the South: Where Is All the Wood? *Journal of Forestry* 93(7): 16–20.

Daugherty, Arthur B. 1995. *Major Uses of Land in the United States, 1992.* Agricultural Economic Report no. 723. Washington, DC: U.S. Department of Agriculture, Natural Resources and Environment Division, Economic Research Service.

DeForest, Christopher E., Thomas G. Harris Jr., Frederick W. Cubbage, and Arthur C. Nelson. 1991. Timberland Downtown?: Southern Forest Resources Along the Urban-rural Continuum. Pages 137–138 in *Proceedings of a Symposium, Ecological Land Classification: Applications to Identify the Productive Poten-*

tial of Southern Forests. General Technical Report SE-68. Asheville, NC: U.S. Department of Agriculture, USDA Forest Service, Southeast Forest Experiment Station.

DellaSala, Dominick A., David M. Olson, Sam E. Barth, Saundra L. Crane, and Steve A. Primm. 1995. Forest Health: Moving beyond the Rhetoric to Restore Healthy Landscapes in the Inland Northwest. *Wildlife Society Bulletin* 23(3): 346–356.

Dietrich, William. 1999. How Progress Ate America. *American Forests* 105(3): 24–29.

Edmonds, Robert. 1997. Program leader, Extension Service, University of New Hampshire, Durham. Telephone. interview, January 3, 1997.

Egan, A. F., and S. B. Jones. 1993. Do Landowner Practices Reflect Beliefs? *Journal of Forestry* 91(10): 39–45.

Ellefson, Paul V., Anthony S. Cheng, and Robert J. Moulton. 1995. *Regulation of Private Forestry Practices by State Governments.* Station Bulletin 605–1995. St. Paul: University of Minnesota, Minnesota Agricultural Experiment Station.

Ewel, John J., et al. 1999. Deliberate Introductions of Species: Research Needs. *Bioscience* 49(8): 619–630.

Flather, Curtis H., Stephan J. Brady, and Michael S. Knowles. 1999. *Wildlife Resource Trends in the United States: A Technical Document Supporting the 2000 USDA Forest Service RPA Assessment.* General Technical Report RMRS-GTR-33. Fort Collins, CO: U.S. Department of Agriculture, USDA Forest Service, Rocky Mountain Research Station.

Flather, Curtis H., Linda A. Joyce, and Carol A. Bloomgarden. 1994. *Species Endangerment Patterns in the United States.* General Technical Report RM-241. Fort Collins, CO: U.S. Department of Agriculture, USDA Forest Service, Rocky Mountain Research Station.

Fletcher, Richard A., and A. Scott Reed. 1996. Extending Forest Management with Volunteers: The Master Woodlands Manager Project. Pages 69–81 in *Proceedings: Symposium on Non-Industrial Private Forests: Learning from the Past, Prospects for the Future.* Washington, DC, February 18–20, 1996.

Freisen, Lyle E., Paul F. J. Eagles, and R. J. Mackay. 1995. Effects of Residential Development on Forest-dwelling Neotropical Migrant Songbirds. *Conservation Biology* 9(6): 1408–1414.

Fricker, Mary. 1999. Conservation Easements: Popular Way to Protect Land, But Much Work Lies Ahead. *Press Democrat,* April 4 ,1999, page E1.

Gilmore, Dan, Ted Beauvais, and Elizabeth Zucker. 1999. Federal Dollars to Protect Forests: Using the Forest Legacy Program. Session 6K. In *National Land Trust Rally '99,* Snowmass, CO, October 14–17, 1999. Washington, DC: Land Trust Alliance.

Gordon, John, et al. 1993. *An Assessment of Indian Forests and Forest Management in the United States.* Report of the IFMAT. Portland, OR: Intertribal Timber Council.

Gordon, John, Jerry F. Franklin, K. Norman Johnson, Dave Patton, Jim Sedell, John Sessions, and Ed Williston. 1997. An Independent Report on Tribal Forestry: Redefining the Government Role. *Journal of Forestry* 95(11): 10–14.

Gray, Gerald J., and Paul V. Ellefson. 1987. *Statewide Forest Resource Planning: The Effects of First-Generation Programs.* Miscellaneous Publication 20-1987. St. Paul: University of Minnesota, Agricultural Experiment Station.

Greene, John, Tamara Cushing, Steve Bullard, and Ted Beauvais. 1999. Effect of the Federal Estate Tax on Nonindustrial Forestland Owners. In *Conference Proceedings, Keep America Growing: Balancing Working Lands and Development.* Philadelphia, June 6–9. Compact disk.

Greene, John L., and William C. Siegel. 1994. *The Status and Impact of State and Local Regulation on Private Timber Supply.* General Technical Report RM-255. Fort Collins, CO: U.S. Department of Agriculture, USDA Forest Service, Rocky Mountain Research Station.

Guldin, James M., and T. Bently Wigley. 1998. Intensive Management: Can the South Really Live without It? In: *Proceedings of the 63rd North American Wildlife and Natural Resources Conference,* 1998.

Guldin, Richard W. 1989. *An Analysis of the Water Situation in the United States: 1989–2040.* Technical Document Supporting the 1989 USDA Forest Service RPA Assessment. General Technical Report RM-177. Fort Collins, CO: U.S. Department of Agriculture, USDA Forest Service, Rocky Mountain Forest and Range Experiment Station.

Hacker, Jan J., and Melvin J. Baughman. 1995. Designing Innovative Funding for State Forestry Programs. *Journal of Forestry* 93(8): 10–13.

Harris, Larry D. 1984. *The Fragmented Forest Island Biogeography Theory and the Preservation of Biotic Diversity.* Chicago: University of Chicago Press.

Haynes, Richard W. 1990. *An Analysis of the Timber Situation in the United States: 1989–2040.* Technical Document Supporting the 1989 USDA Forest Service RPA Assessment. General Technical Report RM-199. Fort Collins, CO: U.S. Department of Agriculture, USDA Forest Service, Rocky Mountain Forest and Range Experiment Station.

———. 1999. Program manager for Social and Economic Values, U.S. Department of Agriculture. Forestry Science Lab. Telephone interview and timber harvest data of the United States, December 8, 1999.

———. 2001. *The 2000 RPA Timber Assessment: An Analysis of the Timber Situation in the United States, 1996 to 2050* (Draft January 10, 2001). USDA Forest Service. http://www.fs.fed.us/pnw/sev/rpa/index.htm

Haynes, Richard W., Darius M. Adams, and John R. Mills. 1995. *The 1993 RPA Timber Assessment Update.* General Technical Report RM-259. Fort Collins, CO: U.S. Department of Agriculture, USDA Forest Service, Rocky Mountain Forest and Range Experiment Station.

Hodge, Sandra S. 1999. Research assistant professor, University of Missouri. Telephone interview, Wednesday, August 18, 1999.

Hodge, Sandra S., and L. Southhard. 1992. A Profile of Virginia NIPF Landowners: Results of a 1991 Survey. *Virginia Forests* 47(4): 7–9, 11.

Intertribal Timber Council (ITC). 1993. *An Assessment of Indian Forests and Forest Management in the United States.* Portland, OR: Intertribal Timber Council.

Johnson, Rebecca L., Ralph J. Alig, Eric Moore, and Robert J. Moulton. 1997. NIPF: Landowners' View of Regulation. *Journal of Forestry* 95(1): 23–28.

Jones, Stephan B., Albert E. Luloff, and James C. Finley. 1995. Another Look at NIPFs, Facing Our "Myths." *Journal of Forestry* 93(9): 41–44.

Keisling, Phil, ed. 1999. *Oregon Blue Book 1999–2000.* Salem, OR: Office of the Secretary of State.

Kline, Jeffrey D., and Ralph J. Alig. 1999. Does Land Use Planning Slow the Conversion of Forest and Farm Lands? *Growth and Change* 30: 3–22.

Kohm, Kathryn A., and Jerry F. Franklin, eds. 1997. *Creating a Forestry for the 21ˢᵗ Century: The Science of Ecosystem Management.* Washington, DC: Island Press.

Land Trust Alliance. 1998. URL: *http://ww.lta.org*

Langer, Linda L., and Curtis H. Flather. 1994. *Biological Diversity: Status and Trends in the United States.* General Technical Report RM-244. Fort Collins, CO: U.S. Department of Agriculture, USDA Forest Service, Rocky Mountain Forest and Range Experiment Station.

Lerner, Steve, and William Poole. 1999. The Economic Benefits of Parks and Open Space. San Francisco: The Trust for Public Land.

Likens, G. E., C. T. Driscoll, and D. C. Buso. 1996. Long-term Effects of Acid Rain: Response and Recovery of a Forest Ecosystem. *Science* 272 (April 12, 1996).

Liu, Rei, and John A. Scrivani. 1997. *Virginia Forest Land Assessment.* Phase One Project Report 1997. Forest Resource Assessment. Charlottesville: Virginia Department of Forestry.

Mac, Michael J., et al. 1998. *Status and Trends of the Nation's Biological Resources.* 2 vols. Reston, VA: U.S. Department of the Interior, U.S. Geological Survey.

McColly, Robert. 1996. Consulting Foresters Perspective. In M. J. Baughman (ed)., *Proceedings: Symposium on Nonindustrial Private Forests: Learning from the Past, Prospects for the Future.* St. Paul: Minnesota Extension Service.

Miller Freeman. 1998. *Lumber and Panel: North American Fact Book.* San Francisco: Miller Freeman Publishing.

Minnesota Forest Resources Council. 1999. *Biennial Report to the Governor and Legislature on Sustainable Forest Resources Act Implementation, 1997–1998.* St. Paul: Minnesota Forest Resources Council.

Mladenoff, David J., and John Pastor. 1993. Sustainable Forest Ecosystems in the Northern Hardwood and Conifer Forest Region: Concepts and Management. Pages 145–180 in *Defining Sustainable Forestry.* Washington, DC: Island Press.

Morishima, Gary. 1997. Indian Forestry: From Paternalism to Self-Determination. *Journal of Forestry* 95(11): 4–9.

Moulton, Robert J., Felicia Lockhart, and Jeralyn D. Snellgrove. 1995. *Tree Planting in the United States, 1995.* Washington, DC: U.S. Department of Agriculture, USDA Forest Service, Cooperative Extension Service.

National Research Council. 1998. *Forested Landscapes in Perspective: Prospects and Opportunities for Sustainable Management of America's Nonfederal Forests.* Committee on Prospects and Opportunities for Sustainable Management of America's Nonfederal Forests. Board on Agriculture. Washington, DC: National Academy Press.

New Hampshire Forest Sustainability Standards Work Team. 1997. *Good Forestry in the Granite State: Recommended Voluntary Forest Management Practices for New Hampshire.* Concord: Society for the Protection of New Hampshire Forests.

Newman, David H., and David N. Wear. 1993. The Production Economics of Private Forestry: A Comparison of Industrial and Nonindustrial Forest Owners. *The American Journal of Agriculture Economics* 75(3): 674–695.

Nilsson, Sten, Ralph Colberg, Robert Hagler, and Peter Woodbridge. 1999. *How Sustainable Are North American Wood Supplies?* Interim Report IR-99-003. January 1999. Laxenburg, Austria: International Institute for Applied Systems Analysis.

Noss, Reed F. 1983. A Regional Landscape Approach to Maintaining Diversity. *Bioscience* 33: 700–706.

———. 1987. Protecting Natural Areas in Fragmented Landscapes. *Natural Areas Journal* 7: 2–13.

Noss, Reed F., and Allen Y. Cooperrider. 1994. *Saving Nature's Legacy.* Washington, DC: Island Press.

Noss, Reed F., and Robert L. Peters. 1995. *Endangered Ecosystems: A Status Report on America's Vanishing Habitat and Wildlife.* Washington, DC: Defenders of Wildlife.

Odum, Eugene. 1998. *Ecological Vignettes.* Amsterdam, BV: Hardwood Academic Publishers.

Oliver, Chadwick, et al. 1997. *Report on Forest Health of the United States: A Panel Chartered by Congressman Charles Taylor.* Reprinted by CINTRAFOR, RE43A. Seattle: University of Washington.

Oregon Department of Forestry. 1997. *Oregon Forests Report 1997.* Salem: Oregon Department of Forestry.

Owen, Carlton. 1999. Vice president–Forest Policy, Champion International Corporation. Telephone interview, Tuesday, August 31, 1999.

Pacific Forest Trust. 1999. *Changes in On-site Carbon Stores Resulting from Transitions in Silviculture on Forests Managed by MacMillian Bloedel: A Report for the World Resources Institute.* Boonville, CA: The Pacific Forest Trust.

Park, Chris C. 1987. *Acid Rain: Rhetoric and Reality.* New York: Methuen & Co.

Parks, Peter J., and Brian C. Murrary. 1994. Land Attributes and Land Allocation: Nonindustrial Forest Use in the Pacific Northwest. *Forest Science* 40(3): 558–575.

Powell, Douglas S., Joanne L. Faulkner, David R. Darr, Zhiliang Zhu, and Douglas W. MacCleery 1993. Revised 1994. *Forest Resources of the United States, 1992.* General Technical Report RM-234. Fort Collins, CO: U.S. Department of Agriculture, USDA Forest Service, Rocky Mountain Forest and Range Experiment Station.

Rathke, David M., and Melvin J. Baughman. 1996. Influencing Nonindustrial Private Forest Management through the Property Tax System. *Northern Journal of Applied Forestry* 13(1): 30–36.

Ricketts, Taylor H., Eric Dinerston, David M. Olsen, Colby Loucks, William Eichbaum, Dominick DellaSala, Kevin Kavanagh, Hedao Prashart, Patrick T. Hurley, Karen M. Carrey, Robin Abell, and Steven Walter. 1999. *Terrestrial Ecoregions of North America: A Conservation Assessment.* Washington, DC: Island Press.

Romm, Jeff, Raul Tuazon, Court Washburn, Judy Bendix, and James Rinehart. 1983. *The Non-Industrial Forestland Owners of Northern California.* Berkeley: University of California, Department of Forestry and Resource Management.

Rosenberg, Kenneth V., and Martin G. Raphael. 1986. Effects of Forest Fragmentation on Vertebrates in Douglas-fir Forests. Pages 263–272 in Jared Verner, Michael L. Marrison, and C. John Ralph (eds.) *Wildlife 2000: Modeling Habitat Relationships of Terrestrial Vertebrates.* Madison: University of Wisconsin Press.

Royer, Jack P., and Robert J. Moulton. 1987. Reforestation Incentives: Tax Incentives and Cost Sharing in the South. *Journal of Forestry* 85(8): 45–47.

Sampson, R. Neil. 1999. Sustaining America's Forests—Where We Are. Presented at the NAPFC/CSREES *Summit on Sustaining America Forests: The Role of Research, Education, and Extension.* Washington, DC, February 22, 1999.

Sampson, R. Neil, and Lester DeCoster. 1996. *Mini-Survey Forestry Program Providers and Participants.* Washington, DC: American Forests, Forest Policy Center.

———. 1997. *Public Programs for Private Forestry: A Reader on Programs and Options.* Washington, DC: American Forests.

———. 1998. *Forest Health in the United States.* Washington, DC: American Forests.

Saunders, Denis A., Richard J. Hobbs, and Chris R. Margules. 1991. Biological Consequences of Ecosystem Fragmentation: A Review. *Conservation Biology* 5: 18–32.

Sharp, Brian E. 1996. Avian Population Trends in the Pacific Northwest. *Bird Populations* 3: 26–45.

Smith, W. Brad, and Raymond Sheffield. 2000. A Brief Overview of the Forest Resources of the United States, 1997. USDA Forest Service. http://fia.fs.fed.us/rpa.htm

Starkman, Dean. 1999. Southern Chip Mills Lead to Forestry Flap. *Wall Street Journal,* vol. CXLI no. 61, Monday, September 27, 1999, front page and A14.

Turner, David P., Greg J. Koeper, Mark E. Harmon, and Jeffrey J. Lee. 1995. Carbon Sequestration by Forests of the United States: Current Status and Projections to Year 2040. *Tellus.* 47B: 232–239.

U.S. Department of Agriculture, USDA Forest Service. 1999. *U.S. Department of Agriculture Talking Points on Invasive Species (revised).* Washington, DC: National Council on Invasive Species.

————. 2000. *Resource Planning Assessment Database.* http://fia.fs.fed.us

U.S. Department of Agriculture, Natural Resources Conservation Service. 1994. *National Resources Inventory 1992.* Data tables obtained Monday, August 30, 1999. http://www.nhq.nrcs.usda.gov/NRI/1997/

U.S. Department of Agriculture, Natural Resources Conservation Service. 1999. Summary Report 1997 National Resources Inventory (Revised December 2000). http://www.nhq.nrcs.usda.gov/NRI/1997/

U.S. Department of Energy. Energy Information Agency. 1997. *Emissions of Greenhouse Gases in the United States 1997.* DOE/EIA-0573(97). Washington, DC: U.S. Government Printing Office.

U.S. Department of the Interior. U.S. Geological Survey. 1998. *Status and Trends of the Nation's Biological Resources.* Vols. 1 and 2. Washington, DC: U.S. Government Printing Office.

U.S. Environmental Protection Agency. Nonpoint Source Control Branch. 1997a. *Nonpoint Source Pollution: The Nation's Largest Water Quality Problem.* Pointer No. 1. EPA841-F-96-004A. URL: http://www.epa. gov/OWOW/NPS/facts/point1.html. Information obtained 12/3/99.

————. Nonpoint Source Control Branch. 1997b. *Managing Nonpoint Source Pollution from Forestry.* Pointer No. 8. EPA841-F-96-004H. URL: http://www.epa.gov/OWOW/NPS/facts/point8.html. Information obtained 12/3/99.

————. *Unified Watershed Assessment 1998.* URL: http://www.epa.gov/owowwtr1/tmdl/

U.S. Fish and Wildlife Service. 1999. *Habitat Conservation Plans.* Environmental Conservation Online System (ECOS). December 2, 1999. URL: http://ecos.fws.gov/hcp_report/

U.S. General Accounting Office. Resources, Community, and Economics Development Division. 1994. *Endangered Species Act: Information on Species Protection on Nonfederal Lands.* GAO-RCED-95-16. Washington, DC: U.S. Government Printing Office.

Wear, David N. 1993. *Private Forest Investment and Softwood Production in U.S. South.* General Technical Report RM-237. Fort Collins, CO: U.S. Department of Agriculture, USDA Forest Service, Rocky Mountain Forest and Range Experiment Station.

————. 1996. *Forest Management and Timber Production in the U.S. South.* SCFER Working Paper no. 82, 40. Research Triangle Park, NC: Southeastern Center for Forest Economics Research.

Wear, David N., Rei Liu, J. Michael Foreman, and Raymond M. Sheffield. 1999. The Effects of Population Growth on Timber Management and Inventories in Virginia. *Forest Ecology and Management* 118(1999): 107–115.

Wear, David N., Robert Abt, and Robert Mangold. 1998. People, Space and Time: Factors That Will Govern Forest Sustainability. In *Proceedings of the 63rd North American Wildlife and Natural Resources Conference,* 1998.

Westbrooks, Randy G. 1998. *Invasive Plants, Changing the Landscape of America: Fact Book.* Washington, DC: Federal Interagency Committee for the Management of Noxious and Exotic Weeds.

Wilcox, B. A., and D. D. Murphy. 1985. Conservation Strategy: The Effects of Fragmentation on Extinction. *American Naturalist* 125: 879–887.

Zheng, Daolan. 1989. "Factors Affecting Conversion of Agricultural Land to Residential Use in Kittias County, WA." Master's thesis, Central Washington University, Ellensburg.

Zheng, Daolan, and Ralph J. Alig. 1999. *Changes in the Non-Federal Land Base Involving Forestry in Western Oregon, 1961–64.* PNW-RP-518. Portland, OR: U.S. Department of Agriculture USDA, Forest Service, Pacific Northwest Research Station.

Zimmerman, Elliot. 1976. *A Historical Summary of State and Private Forestry in the U.S. Forest Service.* Washington, DC: USDA Forest Service, State and Private Forestry.

Index

About the Authors

Constance Best is the cofounder and managing director of the Pacific Forest Trust. A national leader in conservation of productive forestlands, Ms. Best has promoted an understanding of the synergy between conservation and commerce. In addition to other publications, she is the principal author of *Capital Markets and Sustainable Forestry: Opportunities for Investment,* a report for the John D. and Catherine T. MacArthur Foundation. Ms. Best is an accomplished entrepreneur and consumer products marketer, having cofounded and managed the first natural soda business in America, Soho Natural Sodas (sold in 1989 to Seagrams). She has served on the boards of Ecotrust, The Land Trust Alliance, the Anderson Valley Land Trust, and the Investor's Circle (an association of socially responsible venture capital investors).

Laurie A. Wayburn is the cofounder and president of the Pacific Forest Trust. Serving in both the United Nations Environment Program and UNESCO's Man and Biosphere Program, Ms. Wayburn advanced the twin goals of environmental protection and economic development in Africa, Latin America, and Europe. She has served on boards or key committees of numerous organizations to further land conservations and stewardship, including the Land Trust Alliance, the Seventh American Forest Congress, the University of California Center for Forestry, the Oregon Board of Forestry Incentives Group, the California Forestland Incentives Task Force, the United States Man and Biosphere Committee for Biosphere Reserves, the Society of American Foresters Certification Task Force, and the Compton Foundation.